Trends in Mathematics

Trends in Mathematics is a series devoted to the publication of volumes arising from conferences and lecture series focusing on a particular topic from any area of mathematics. Its aim is to make current developments available to the community as rapidly as possible without compromise to quality and to archive these for reference.

Proposals for volumes can be submitted using the Online Book Project Submission Form at our website www.birkhauser-science.com.

Material submitted for publication must be screened and prepared as follows:

All contributions should undergo a reviewing process similar to that carried out by journals and be checked for correct use of language which, as a rule, is English. Articles without proofs, or which do not contain any significantly new results, should be rejected. High quality survey papers, however, are welcome.

We expect the organizers to deliver manuscripts in a form that is essentially ready for direct reproduction. Any version of TEX is acceptable, but the entire collection of files must be in one particular dialect of TEX and unified according to simple instructions available from Birkhäuser.

Furthermore, in order to guarantee the timely appearance of the proceedings it is essential that the final version of the entire material be submitted no later than one year after the conference.

More information about this series at http://www.springer.com/series/4961

Jörn Steuding
Editor

Diophantine Analysis

Course Notes from a Summer School

With Contribution by
Sanda Bujačić
Alan Filipin
Simon Kristensen
Tapani Matala-aho
Nicola M.R. Oswald

Editor
Jörn Steuding
Institut für Mathematik
Universität Würzburg
Würzburg, Bayern
Germany

ISSN 2297-0215 ISSN 2297-024X (electronic)
Trends in Mathematics
ISBN 978-3-319-84020-8 ISBN 978-3-319-48817-2 (eBook)
DOI 10.1007/978-3-319-48817-2

Mathematics Subject Classification (2010): 46-06, 47-06, 28-06, 06-06

Printed on acid-free paper

This book is published under the trade name Birkhäuser
The registered company is Springer International Publishing AG
The registered company address is: Gewerbestrasse 11, 6330 Cham, Switzerland
(www.birkhauser-science.com)

UNIVERSITÄT WÜRZBURG

DIOPHANTINE ANALYSIS
Summer School, 21- 26 July 2014

The Department of Mathematics at the University of Würzburg invites to a summer school on number theory with focus on diophantine analysis!

A Geometric Face of Diophantine Analysis

Tapani Matala-aho
University of Oulu

Linear Forms in Logarithms

Sanda Bujačić
University of Rijeka

Simon Kristensen
Aarhus University

Metric Diophantine Approximation - From Continued Fractions to Fractals

Nicola Oswald
Universität Würzburg

Historical Face of Number Theory(ists) at the turn of the 19th Century

The Summer School is addressed to Master and PhD-Students and all those interested in these four faces of Number Theory. Deadline for application will be May 16, 2014. For registration and further information including scholarships please visit our webpage
http://www.mathematik.uni-wuerzburg.de/~steuding/start2014.htm.

We are looking forward to seeing you in Würzburg! Jörn Steuding, Rasa Steuding, Teerapat Srichan, Nicola Oswald

Preface

Number theory is one of the oldest mathematical disciplines. Often enough number theoretical problems are easy to understand but difficult to solve, typically only by using advanced methods from other mathematical disciplines; a prominent example is Fermat's last theorem. This is probably one of the reasons why number theory is considered to be such an attractive field that has intrigued mathematicians and math lovers for over 2000 years.

The subdiscipline called *Diophantine analysis* may be defined as the combination of the theories of Diophantine approximation and Diophantine equations. In both areas, the nature of numbers plays a central role. For instance, a celebrated theorem of Klaus Roth (worth a Fields Medal) states that, roughly speaking, algebraic numbers cannot be approximated by rationals too well (see Theorem 1.4.21 in Chapter "Linear Forms in Logarithms"). As a consequence of Roth's theorem, one can show that certain cubic equations have only finitely many integer solutions, e.g.

$$aX^3 + bY^3 = c$$

with arbitrary, but fixed nonzero integers a, b, c.

In July 2014, the number theory group of the Department of Mathematics at Würzburg University organised an international summer school on Diophantine analysis. In the frame of this event, about fifty participants, mostly Ph.D. students from all over the world, but also a few local participants, and even undergraduate students, learned in three courses about different topics from Diophantine analysis; a fourth course gave in addition some historical background of some aspects of the early research in this direction.

- Sanda Bujačić (University of Rijeka, Croatia) lectured on *Linear Forms in Logarithms*. Starting with some classical Diophantine approximation theorems, her course focuses on Alan Baker's celebrated results from 1966 on effective lower bounds for the absolute value of a nonzero linear form in logarithms of algebraic numbers (another Fields Medal). His pathbreaking approach goes beyond the classical results and allows many interesting applications. In the course notes, which is joint work with Alan Filipin (University of Zagreb, Croatia), it is shown how to solve the system of two simultaneous Pellian equations (a classical application due to Baker and Davenport), how to find all repdigit Fibonacci numbers (a theorem by Florian Luca from 2000), how to determine the bound of the number of perfect powers in a binary recurrence sequence, etc.

- In the course of Simon Kristensen (Aarhus University, Denmark) on *Metric Diophantine Approximation—From Continued Fractions to Fractals*, the classical Khintchine theorems on metric Diophantine approximation are considered (by studying continued fractions by means of a dynamical system). After a crash course in fractal geometry, the notes outline the major topics of recent research as, e.g., Schmidt's game, the question whether the Cantor middle third set contains an algebraic irrational, and a discussion of badly approximable numbers. One of the highlights is the proof that the set of badly approximable β for which the pair (α, β) satisfies the Littlewood conjecture has Hausdorff dimension one; here, α may even be substituted by any countable set of badly approximable numbers (which is a new result due to Haynes, Jensen, and Kristensen).

- Tapani Matala-aho (University of Oulu, Finland) examines in *A Geometric Face of Diophantine Analysis* the so-called geometry of numbers. Building on the notions of convex sets and lattices as well as Hermann Minkowski's fundamental theorems, classical Diophantine inequalities are deduced. Also, some Diophantine inequalities over complex numbers are discussed. Then, he presents some variations of Siegel's lemma over rational and imaginary quadratic fields supplementing Enrico Bombieri's works. Moreover, building on Wolfgang Schmidt's work, it is proved that the heights of a rational subspace and its orthogonal complement are equal (by the use of Grassmann algebras). His lectures end with a proof of the Bombieri–Vaaler version of Siegel's lemma.

- In her course *Historical Face of Number Theory(ists) at the Turn of the 19th Century*, Nicola Oswald (University of Würzburg, Germany) describes the lives and mathematical works of the famous Adolf Hurwitz and his unknown elder brother Julius around the turn of the nineteenth/twentieth century. A careful discussion of the mathematical diaries of Adolf Hurwitz (or at least some of its aspects) provides an understanding of his mathematics on behalf of historical documents; a particular emphasis is put on his relation to David Hilbert. Moreover, Julius Hurwitz' work on complex continued fractions is investigated and further analysed with modern tools from ergodic theory.

This volume presents the lecture notes of these four summer school courses (some of them with additional material). Each of these notes serves as an essentially self-contained introduction. (Of course, a background in number theory might be useful.) Altogether, the reader gets a thorough impression of Diophantine analysis by its central results, relevant applications, and big open problems. The notes are complemented with many references and an extensive register which makes it easy to navigate through the book.

The authors and the editor are grateful to the anonymous referees for their valuable suggestions and remarks that have improved the book. They also thank Springer for making the collection of lecture notes a book and, in particular, Clemens Heine for his encouragement.

When turning the pages, it is impressive to see how one can approach frontiers of current research in this direction of number theory quickly by only elementary and basic analytic methods. We wish to take much pleasure in reading.

Würzburg, Germany   Jörn Steuding
July 2016

Contents

Linear Forms in Logarithms

Sanda Bujačić and Alan Filipin

2000 Mathematics Subject Classification 11J13 · 11J81 · 11J86 · 11K60

Introduction

These lecture notes cover the course *Linear Forms in Logarithms* created for the Summer School *Diophantine Analysis* organized in Würzburg, Germany in 2014. The notes intend to be an introduction to Diophantine approximation and linear forms in logarithms.

We begin with the theory of Diophantine approximation which has an extremely important application in the study of Diophantine equations. One of the main topics is the question how well a given real number α can be approximated by rational numbers. By placing certain constraints on the rational numbers used in the approximation, we are able to classify the real number α as either a *rational* or an *irrational* number, or as an *algebraic* or a *transcendental* number. Diophantine approximation and transcendence theory are very close areas that share many theorems and methods which will be useful in the second part of these lecture notes.

There, we introduce linear forms in logarithms and provide lower bounds for linear forms in logarithms of algebraic numbers due to Alan Baker, one of the most famous mathematician in this field of mathematics. Baker was awarded the Fields Medal in 1970 because of his profound and significant contributions to number theory. To

S. Bujačić (✉)
Department of Mathematics, University of Rijeka, Radmile Matejčić 2,
51 000 Rijeka, Croatia
e-mail: sbujacic@math.uniri.hr

A. Filipin
Faculty of Civil Engineering, University of Zagreb, Fra Andrije Kačića-Miošića 26,
10 000 Zagreb, Croatia
e-mail: filipin@grad.hr

© Springer International Publishing AG 2016
J. Steuding (ed.), *Diophantine Analysis*, Trends in Mathematics,
DOI 10.1007/978-3-319-48817-2_1

1

illustrate the importance of his machinery, many useful and interesting applications of the introduced concepts are presented.

1 Diophantine Approximation

Rational numbers are in every interval of the real line, no matter how small that interval is because the set of rational numbers is *dense* in the set of real numbers. As a consequence, for any given real number α, we are able to find a rational number as close as we like to α.

Approximating a real number by rational numbers helps us to better understand the set of real numbers and gives us a surprising insight in the properties of real numbers. By placing certain constraints on the rational numbers used in the approximation of the real number α, properties of the real number α can be observed such that classify it as either a *rational* or an *irrational* number, or as an *algebraic* or a *transcendental* number.

As rational numbers approach a fixed real number, their denominators grow arbitrarily large. We study how closely real numbers can be approximated by rational numbers that have a fixed bound on the growth of their denominators.

1.1 Dirichlet's Theorem

One of the main questions in Diophantine approximation is whether there exists *any* rational number $\frac{p}{q}$ satisfying the inequality

$$\left| \frac{a}{b} - \frac{p}{q} \right| < \frac{1}{q^2}.$$

The affirmative answer follows from the fundamental result due to Dirichlet[1] on rational approximation of real numbers.

Theorem 1.1.1 (Dirichlet 1842) *Let α be a real number and n a positive integer. There exists a rational number $\frac{p}{q}$, $0 < q \leq n$, satisfying the inequality*

$$\left| \alpha - \frac{p}{q} \right| \leq \frac{1}{(n+1)q}. \tag{1.1}$$

Proof For $n = 1$, $\frac{p}{q} = \frac{\lfloor \alpha \rfloor}{1}$ or $\frac{p}{q} = \frac{\lfloor \alpha \rfloor + 1}{1}$ satisfies

---

[1] Peter Gustav Lejeune Dirichlet (1805–1859), a German mathematician.

$$\left| \alpha - \frac{p}{q} \right| \le \frac{1}{2},$$

where $\lfloor \; \rfloor$ stands for floor function. For $n \ge 2$, we consider $n + 2$ elements

$$0, \; \alpha - \lfloor \alpha \rfloor, \; 2\alpha - \lfloor 2\alpha \rfloor, \; \ldots, \; n\alpha - \lfloor n\alpha \rfloor, \; 1,$$

of the interval $[0, 1]$.

Assume first that these elements are distinct as they are in the case α is an irrational number. The interval $[0, 1]$ can be subdivided into $n + 1$ subintervals of the length $\frac{1}{n+1}$ and the pigeonhole principle guarantees that two of these numbers differ in absolute value by at most $\frac{1}{n+1}$. If one of the numbers is 0 and the other is $i\alpha - \lfloor i\alpha \rfloor$, then $i \le n$, $|i\alpha - \lfloor i\alpha \rfloor| < \frac{1}{n+1}$ and

$$\left| \alpha - \frac{\lfloor i\alpha \rfloor}{i} \right| \le \frac{1}{(n+1)i}.$$

After $\frac{\lfloor i\alpha \rfloor}{i}$ is reduced to lowest terms $\frac{p}{q}$, the rational number $\frac{p}{q}$ satisfies (1.1). Similarly, if the two numbers in question are $j\alpha - \lfloor j\alpha \rfloor$ and 1, then $j \le n$ and reducing $\frac{\lfloor j\alpha \rfloor + 1}{j}$ to lowest terms $\frac{p}{q}$, we have that $\frac{p}{q}$ satisfies (1.1). Finally, if the two numbers are $i\alpha - \lfloor i\alpha \rfloor$ and $j\alpha - \lfloor j\alpha \rfloor$ with $i < j$, then

$$|j\alpha - \lfloor j\alpha \rfloor - (i\alpha - \lfloor i\alpha \rfloor)| = |(j-i)\alpha - (\lfloor j\alpha \rfloor - \lfloor i\alpha \rfloor)| \le \frac{1}{n+1}.$$

Consequently, $j - i < n$ and

$$\left| \alpha - \frac{\lfloor j\alpha \rfloor - \lfloor i\alpha \rfloor}{j - i} \right| \le \frac{1}{(n+1)(j-i)}.$$

Thus, after $\frac{\lfloor j\alpha \rfloor - \lfloor i\alpha \rfloor}{j-i}$ is reduced to lowest terms $\frac{p}{q}$, the rational number $\frac{p}{q}$ satisfies (1.1).

If the $n + 2$ numbers from the beginning are not distinct, then α itself is a rational number with denominator at most n. In this case, there exist $i < j$ so that α is equal to one of the following fractions

$$\frac{\lfloor i\alpha \rfloor}{i}, \; \frac{\lfloor j\alpha \rfloor - \lfloor i\alpha \rfloor}{j - i}$$

reduced to lowest terms. If the numbers are not distinct, the required inequality (1.1) is trivially satisfied by α itself.

Corollary 1.1.2 *For α irrational, there exist infinitely many relatively prime numbers p, q such that*

$$\left| \alpha - \frac{p}{q} \right| < \frac{1}{q^2}. \tag{1.2}$$

Proof Suppose there are only finitely many rationals

$$\frac{p_1}{q_1}, \frac{p_2}{q_2}, \ldots, \frac{p_k}{q_k}$$

satisfying (1.2). In this case,

$$\left| \alpha - \frac{p_i}{q_i} \right| > 0$$

for $1 \leq i \leq k$. Consequently, since α is irrational, there exists a positive integer n such that the inequality

$$\left| \alpha - \frac{p_i}{q_i} \right| > \frac{1}{n+1}$$

holds for $1 \leq i \leq k$. However, this contradicts Dirichlet's Theorem 1.1.1 which asserts that, for this n, there exists a rational number $\frac{p}{q}$ with $q \leq n$ such that

$$\left| \alpha - \frac{p}{q} \right| \leq \frac{1}{(n+1)q} < \frac{1}{q^2}.$$

Remark 1.1.3 Corollary 1.1.2 does not hold for α rational. Assume $\alpha = \frac{u}{v}$. For $\frac{p}{q} \neq \alpha$, we have

$$\left| \alpha - \frac{p}{q} \right| = \left| \frac{u}{v} - \frac{p}{q} \right| \geq \frac{1}{vq},$$

so (1.1) implies $q < v$, hence the inequality (1.1) can be satisfied only for finitely many relatively prime integers p, q.

Corollary 1.1.4 *A real number α is irrational if and only if there are infinitely many rational numbers $\frac{p}{q}$ such that*

$$\left| \alpha - \frac{p}{q} \right| \leq \frac{1}{q^2}.$$

A proof can be found in [44]. It appears that irrational numbers can be distinguished from rational numbers by the fact that they can be approximated by infinitely many rational numbers $\frac{p}{q}$ with an error less than $\frac{1}{q^2}$.

We may ask for the best possible value $C > 0$ such that the statement of the Corollary 1.1.4 holds with $\frac{1}{Cq^2}$ in place of $\frac{1}{q^2}$? The answer will be given in Hurwitz's[2] Theorem 1.3.5 which characterizes the best Dirichlet-type inequality.

---

[2] Adolf Hurwitz (1859–1919), a German mathematician.

1.2 Continued Fractions

Definition 1.2.1 • An *infinite generalized continued fraction* is an expression of the form

$$a_0 + \cfrac{b_1}{a_1 + \cfrac{b_2}{a_2 + \cfrac{b_3}{\ddots}}}, \qquad (1.3)$$

where $a_0, a_1, a_2, \ldots$ and $b_1, b_2, \ldots$ are either rational, real or complex numbers or functions of such variables.

• For $b_i = 1$, $i \in \mathbb{N}$, (1.3) is called an *infinite simple continued fraction*. Its abbreviated notation is

$$[a_0, a_1, a_2, \ldots].$$

• An expression

$$[a_0, a_1, \ldots, a_n] = a_0 + \cfrac{1}{a_1 + \cfrac{1}{a_2 + \cdots + \cfrac{1}{a_{n-1} + \cfrac{1}{a_n}}}}$$

is called a *finite simple continued fraction* with $a_i \geq 1$, $i = 1, \ldots, n$ and $a_n \geq 2$ integers that are called *partial quotients*. The rational numbers

$$\frac{p_0}{q_0} = [a_0], \ \frac{p_1}{q_1} = [a_0, a_1], \ \frac{p_2}{q_2} = [a_0, a_1, a_2], \ \ldots, \ \frac{p_n}{q_n} = [a_0, a_1, a_2, \ldots, a_n]$$

are called the *convergents* of $\frac{p}{q}$ and n is its *length*.

Remark 1.2.2 We set $p_{-2} := 0$, $p_{-1} := 1$, $q_{-2} := 1$, $q_{-1} = 0$.

Lemma 1.2.3 (Law of formation of the convergents) *For $n \geq 0$,*

$$p_n = a_n p_{n-1} + p_{n-2}, \quad q_n = a_n q_{n-1} + q_{n-2}. \qquad (1.4)$$

Proof The equalities (1.4) are satisfied for $n = 0$. We assume that (1.4) is satisfied for $n - 1$, i.e.,

$$p_{n-1} = a_{n-1} p_{n-2} + p_{n-3}, \quad q_{n-1} = a_{n-1} q_{n-2} + q_{n-3}.$$

Then we get

$$\frac{p_n}{q_n} = [a_0, a_1, \ldots, a_{n-1} + 1/a_n]$$

$$= \frac{(a_{n-1} + \frac{1}{a_n})p_{n-1} + p_{n-2}}{(a_{n-1} + \frac{1}{a_n})q_{n-1} + q_{n-2}}$$

$$= \frac{(a_n a_{n-1} + 1)p_{n-2} + a_n p_{n-1}}{(a_n a_{n-1} + 1)q_{n-2} + a_n q_{n-1}} = \frac{a_n p_{n-1} + p_{n-2}}{a_n q_{n-1} + q_{n-2}}.$$

Lemma 1.2.4 *For* $n \geq -1$,

$$q_n p_{n-1} - p_n q_{n-1} = (-1)^n. \tag{1.5}$$

Proof Let $n = -1$. We observe that $q_{-1}p_{-2} - p_{-1}q_{-2} = (-1)^{-1}$. We assume that (1.5) is satisfied for $n - 1$. Using Lemma 1.2.3, we find

$$q_n p_{n-1} - p_n q_{n-1} = (a_n q_{n-1} + q_{n-2})p_{n-1} - (a_n p_{n-1} + p_{n-2})q_{n-1} =$$
$$= -(q_{n-1}p_{n-2} - p_{n-1}q_{n-2}) = (-1)^n.$$

Lemma 1.2.5 *For* $n \geq 0$,

$$q_n p_{n-2} - p_n q_{n-2} = (-1)^{n-1} a_n.$$

Proof It follows from Lemmas 1.2.3 and 1.2.4 that

$$q_n p_{n-2} - p_n q_{n-2} = (a_n q_{n-1} + q_{n-2})p_{n-2} - (a_n p_{n-1} + p_{n-2})q_{n-2} =$$
$$= a_n(q_{n-1}p_{n-2} - p_{n-1}q_{n-2}) = (-1)^{n-1} a_n.$$

Theorem 1.2.6 *The convergents $\frac{p_k}{q_k}$ satisfy the following inequalities:*

(i) $\frac{p_0}{q_0} < \frac{p_2}{q_2} < \frac{p_4}{q_4} < \ldots$,

(ii) $\frac{p_1}{q_1} > \frac{p_3}{q_3} > \frac{p_5}{q_5} > \ldots$,

(iii) *for n even and m odd,*

$$\frac{p_n}{q_n} < \frac{p_m}{q_m}.$$

Proof Using Lemma 1.2.5, we find

$$\frac{p_{n-2}}{q_{n-2}} - \frac{p_n}{q_n} = \frac{(-1)^{n-1} a_n}{q_{n-2}q_n}.$$

For $n \geq 2$ and n even, we obtain $\frac{p_{n-2}}{q_{n-2}} < \frac{p_n}{q_n}$, while $\frac{p_{n-2}}{q_{n-2}} > \frac{p_n}{q_n}$, for $n \geq 3$ and n odd.

It remains to prove the last inequality. Let $n < m$. Since $\frac{P_n}{q_n} \leq \frac{P_{m-1}}{q_{m-1}}$, it is sufficient to prove $\frac{P_{m-1}}{q_{m-1}} < \frac{P_m}{q_m}$, which is satisfied by Lemma 1.2.4,

$$q_m P_{m-1} - P_m q_{m-1} = (-1)^m = -1 < 0.$$

The proof for $n > m$ follows analogously.

Lemma 1.2.7 *For an integer a_0 and positive integers $a_1, a_2, \ldots, a_n$, the continued fraction $[a_0, a_1, \ldots, a_n]$ is rational. Conversely, for every rational number $\frac{u}{v}$ there exist $n \geq 0$, an integer a_0 and positive integers $a_1, a_2, \ldots, a_n$ such that*

$$\frac{u}{v} = [a_0, a_1, \ldots, a_n].$$

For $\frac{u}{v} \geq 1$, we have $a_0 \geq 1$.

Proof Using Lemma 1.2.3, we find

$$[a_0, a_1, \ldots, a_n] = [a_0, a_1, \ldots, a_{k-1}, [a_k, a_{k+1}, \ldots, a_n]] = \frac{p_{k-1} r_k + p_{k-2}}{q_{k-1} r_k + q_{k-2}},$$

where $r_k := [a_k, a_{k+1}, \ldots, a_n]$.

Conversely, let $v > 0$ and $\gcd(u, v) = 1$. For $v = 1$, clearly $\frac{u}{v} \in \mathbb{Z}$, hence, setting $a_0 = \frac{u}{v}$, we obtain $\frac{u}{v} = [a_0]$.

If $v > 1$, then there exist $q, r \in \mathbb{Z}$ such that $u = vq + r$, $1 \leq r < v$. We assume $\frac{v}{r} = [a_1, \ldots, a_n]$. Since $\frac{v}{r} > 1$, we conclude that $a_1, \ldots, a_n \in \mathbb{N}$. Hence,

$$\frac{u}{v} = q + \frac{1}{\frac{v}{r}} = q + \frac{1}{[a_1, \ldots, a_n]} = [q, a_1, \ldots, a_n],$$

where $a_0 = q$. Clearly, if $\frac{u}{v} \geq 1$, then $a_0 = q \geq 1$.

Remark 1.2.8 There is a one-to-one correspondence between rational numbers and finite simple continued fractions.

Remark 1.2.9 Let $\frac{u}{v}$ be rational, $\gcd(u, v) = 1$ and $u > v > 0$. It follows from Euclid's algorithm that

$$u = vq_1 + r_1, \quad v = r_1 q_2 + r_2, \quad \ldots, \quad r_{j-1} = r_j q_{j+1},$$

hence

$$\frac{u}{v} = q_1 + \frac{1}{\frac{c}{r_1}} = q_1 + \cfrac{1}{q_2 + \cfrac{r_1}{r_2}} = \cdots = [q_1, q_2, \ldots, q_{j+1}].$$

Example 1.2.10

$$\frac{7}{11} = 0 + \frac{1}{\frac{11}{7}} = 0 + \frac{1}{1 + \frac{4}{7}} = 0 + \frac{1}{1 + \frac{1}{\frac{7}{4}}} =$$

$$= 0 + \frac{1}{1 + \frac{1}{1 + \frac{3}{4}}} = 0 + \frac{1}{1 + \frac{1}{1 + \frac{1}{\frac{4}{3}}}} = 0 + \frac{1}{1 + \frac{1}{1 + \frac{1}{1 + \frac{1}{3}}}},$$

which can be written as

$$\frac{7}{11} = [0, 1, 1, 1, 3].$$

Example 1.2.11 The number e is irrational.

The proof goes as follows: if we assume that e is rational, then it can be represented as

$$e = \sum_{i=0}^{\infty} \frac{1}{i!} = \frac{u}{v}.$$

Let n be an integer such that $n > v$. We define $\frac{p}{q} = \sum_{i=0}^{n} \frac{1}{i!}$, with $q = n!$. Using Remark 1.2.9, we get

$$\frac{1}{v} \leq q \left| e - \frac{p}{q} \right| = n! \sum_{j=1}^{\infty} \frac{1}{(n+j)!}$$

$$< n! \sum_{k=0}^{\infty} \frac{1}{(n+1)!} \frac{1}{(n+1)^k} \leq \frac{1}{n+1} \frac{1}{1 - \frac{1}{n+1}} = \frac{1}{n},$$

a contradiction to $n > v$.

Lemma 1.2.12 *For every integer r, there exist exactly two different simple continued fraction expansions representing r, namely $r = [r]$ and $r = [r - 1, 1]$. For r rational, there exist exactly two different simple continued fraction expansions representing r, namely $[a_0, a_1, \ldots, a_n]$ with $a_n \geq 2$ and $[a_0, a_1, \ldots, a_{n-1}, a_n - 1, 1]$.*

Proof Every $r \in \mathbb{Q}$ can be represented as a finite simple continued fraction, namely

$$r = [a_0, a_1, \ldots, a_n], \quad a_0 \in \mathbb{Z}, \quad a_1, a_2, \ldots, a_n \in \mathbb{N}.$$

If $r \in \mathbb{Z}$, then $n = 0$ and $r = [a_0] = a_0$, resp. $r = [r]$. Otherwise, when $n > 0$, then

$$r = a_0 + \frac{1}{[a_1, \ldots, a_n]},$$

with $[a_1, \ldots, a_n] \geq 1$. In view of $r - a_0 \in \mathbb{Z}$, we conclude that $[a_1, \ldots, a_n] = 1$. Since $a_1 \geq 1$, we get $a_1 = 1, n = 1$, so $a_0 = r - 1$ and $r = [r - 1, 1]$.

Now consider $r = \frac{u}{v}$, $\gcd(u, v) = 1$, $v > 0$. We use induction on v in order to prove the second claim. The case when $v = 1$ has already been considered. If $v > 1$, we get $\frac{u}{v} = a_0 + \frac{u_1}{v}$, where $u_1 < v$. Let $\alpha_1 := \frac{u_1}{v}$. Using the hypothesis of the induction, we conclude that α_1 has two continued fraction expansions $[a_1, a_2, \ldots, a_{n-1}, a_n]$, $a_n \geq 2$ and $[a_1, a_2, \ldots, a_n - 1, 1]$.

Lemma 1.2.13 *Let a_0 be an integer and $a_1, a_2, \ldots,$ be positive integers. Then, the limit $\lim_{n \to \infty} \frac{p_n}{q_n}$ exists and its value is irrational. Conversely, for α irrational, there exist a unique integer a_0 and unique positive integers $a_1, a_2, \ldots$ such that $\alpha = \lim_{n \to \infty} \frac{p_n}{q_n}$.*

Proof In view of $\frac{p_0}{q_0} < \frac{p_2}{q_2} < \cdots < \frac{p_1}{q_1}$, it is clear that both limits

$$\lim_{\substack{n \to \infty \\ n \text{ even}}} \frac{p_n}{q_n} \quad \text{and} \quad \lim_{\substack{n \to \infty \\ n \text{ odd}}} \frac{p_n}{q_n}$$

exist, and it is easily shown that both limits are equal. We put $\alpha = \lim_{n \to \infty} \frac{p_n}{q_n}$ and compute

$$\left| \alpha - \frac{p_n}{q_n} \right| < \left| \frac{p_{n+1}}{q_{n+1}} - \frac{p_n}{q_n} \right| = \frac{1}{q_n q_{n+1}} < \frac{1}{q_n^2}.$$

Since p_n, q_n are relatively prime, there exist infinitely many rational numbers $\frac{p}{q}$ such that $|\alpha - \frac{p}{q}| < \frac{1}{q^2}$, so α is irrational.

Conversely, let α be irrational, $a_0 = \lfloor \alpha \rfloor$, and let $\alpha_1 := a_0 + \frac{1}{\alpha_1}$. We notice that $\alpha_1 > 1$ is irrational. For $k \geq 1$ let $a_k = \lfloor \alpha_k \rfloor$ and $\alpha_k = a_k + \frac{1}{\alpha_{k+1}}$. We observe that $a_k \geq 1$, $\alpha_{k+1} > 1$, and α_{k+1} is irrational. Our goal is to show

$$\alpha = [a_0, a_1, a_2, \ldots].$$

Using Lemmas 1.2.3 and 1.2.4 with $\alpha = [a_0, a_1, \ldots, \alpha_{n+1}]$, we find

$$q_n \alpha - p_n = q_n \frac{\alpha_{n+1} p_n + p_{n-1}}{\alpha_{n+1} q_n + q_{n-1}} - p_n$$

$$= \frac{q_n(\alpha_{n+1} p_n + p_{n-1}) - p_n(\alpha_{n+1} q_n + q_{n-1})}{\alpha_{n+1} q_n + q_{n-1}} = \frac{(-1)^n}{\alpha_{n+1} q_n + q_{n-1}}.$$

Hence,

$$\left| \alpha - \frac{p_n}{q_n} \right| < \frac{1}{q_n^2}, \tag{1.6}$$

which implies $\lim_{n\to\infty} \frac{p_n}{q_n} = \alpha$. Finally, it remains to prove that the integers $a_0, a_1 \geq 1, a_2 \geq 1, \ldots$ are uniquely determined. In view of

$$\alpha = [a_0, a_1, a_2, \ldots] = a_0 + \frac{1}{[a_1, a_2, \ldots]},$$

and $0 \leq \alpha - a_0 < 1$, we find $a_0 = \lfloor \alpha \rfloor$ which implies that a_0 is unique and $\alpha_1 = [a_1, a_2, \ldots]$ is uniquely determined by α. Because $a_1 = \lfloor \alpha_1 \rfloor$, a_1 is unique, etc. This proves the lemma.

1.3 Hurwitz's Theorem

In the sequel we assume α to be irrational. According to (1.6), we conclude that each convergent of α satisfies the inequality

$$\left| \alpha - \frac{p}{q} \right| < \frac{1}{q^2}.$$

Vahlen[3] and Borel[4] have proved the following theorems that deal with approximation properties of two and three consecutive convergents, respectively.

Theorem 1.3.1 (Vahlen 1895, [48]) *Let α be an irrational number and denote by $\frac{p_{n-1}}{q_{n-1}}$, $\frac{p_n}{q_n}$ two consecutive convergents of α. Then, at least one of them satisfies*

$$\left| \alpha - \frac{p}{q} \right| < \frac{1}{2q^2}.$$

Proof We observe that

$$\left| \alpha - \frac{p_n}{q_n} \right| + \left| \alpha - \frac{p_{n-1}}{q_{n-1}} \right| = \left| \frac{p_n}{q_n} - \frac{p_{n-1}}{q_{n-1}} \right| = \frac{1}{q_n q_{n-1}} < \frac{1}{2q_n^2} + \frac{1}{2q_{n-1}^2}.$$

Thus,

$$\left| \alpha - \frac{p_n}{q_n} \right| < \frac{1}{2q_n^2} \quad \text{or} \quad \left| \alpha - \frac{p_{n-1}}{q_{n-1}} \right| < \frac{1}{2q_{n-1}^2}.$$

Theorem 1.3.2 (Borel 1903, [9]) *Let α be an irrational number and denote by $\frac{p_{n-1}}{q_{n-1}}$, $\frac{p_n}{q_n}$, $\frac{p_{n+1}}{q_{n+1}}$ three consecutive convergents of α. Then at least one of them satisfies the inequality*

$$\left| \alpha - \frac{p}{q} \right| < \frac{1}{\sqrt{5}q^2}.$$

---

[3]Theodor Vahlen (1869–1945), an Austrian mathematician.
[4]Émile Borel (1871–1956), a French mathematician.

Proof Let $\alpha = [a_0, a_1, \ldots]$, $\alpha_i = [a_i, a_{i+1}, \ldots]$, $\beta_i = \frac{q_{i-2}}{q_{i-1}}$, $q \geq 1$. It is not difficult to deduce that

$$\left| \alpha - \frac{p_n}{q_n} \right| = \frac{1}{q_n^2(\alpha_{n+1} + \beta_{n+1})}.$$

We show that there does not exist a positive integer n satisfying

$$\alpha_i + \beta_i < \sqrt{5} \tag{1.7}$$

for $i = n - 1, n, n + 1$. Our reasoning is indirect. We assume that (1.7) is satisfied for $i = n - 1, n$. It follows from

$$\alpha_{n-1} = a_{n-1} + \frac{1}{\alpha_n}, \quad \frac{1}{\beta_n} = \frac{q_{n-1}}{q_{n-2}} = a_{n-1} + \frac{q_{n-3}}{q_{n-2}} = a_{n-1} + \beta_{n-1}$$

that

$$\frac{1}{\alpha_n} + \frac{1}{\beta_n} = \alpha_{n-1} + \beta_{n-1} \leq \sqrt{5}.$$

Hence, $1 = \alpha_n \cdot \frac{1}{\alpha_n} \leq (\sqrt{5} - \beta_n)(\sqrt{5} - \frac{1}{\beta_n})$ or, equivalently, $\beta_n^2 - \sqrt{5}\beta_n + 1 \leq 0$ which implies $\beta_n \geq \frac{\sqrt{5}-1}{2}$. For β_n rational, we conclude that $\beta_n > \frac{\sqrt{5}-1}{2}$. Now, if (1.7) is satisfied for $i = n, n + 1$, then again $\beta_{n+1} > \frac{\sqrt{5}-1}{2}$, so we deduce

$$1 \leq a_n = \frac{q_n}{q_{n-1}} - \frac{q_{n-2}}{q_{n-1}} = \frac{1}{\beta_{n+1}} - \beta_n < \frac{2}{\sqrt{5}-1} - \frac{\sqrt{5}-1}{2} < 1,$$

the desired contradiction.

Legendre[5] proved with the following important theorem a converse to the previous results.

Theorem 1.3.3 (Legendre, [32]) *Let p, q be integers such that $q \geq 1$ and*

$$\left| \alpha - \frac{p}{q} \right| < \frac{1}{2q^2}.$$

Then $\frac{p}{q}$ is a convergent of α.

Proof Let $\alpha - \frac{p}{q} = \frac{\varepsilon \nu}{q^2}$ with $0 < \nu < \frac{1}{2}$ and $\varepsilon = \pm 1$. In view of Lemma 1.2.12 there exists a simple continued fraction

$$\frac{p}{q} = [b_0, b_1, \ldots, b_{n-1}]$$

---

[5] Adrien-Marie Legendre (1752–1833), a French matheamtician.

satisfying $(-1)^{n-1} = \varepsilon$. We define ω by

$$\alpha = \frac{\omega p_{n-1} + p_{n-2}}{\omega q_{n-1} + q_{n-2}},$$

such that $\alpha = [b_0, b_1, \ldots, b_{n-1}, \omega]$. Hence,

$$\frac{\varepsilon \nu}{q^2} = \alpha - \frac{p}{q} = \frac{1}{q_{n-1}}(\alpha q_{n-1} - p_{n-1}) = \frac{1}{q_{n-1}} \cdot \frac{(-1)^{n-1}}{\omega q_{n-1} + q_{n-2}},$$

giving $\nu = \frac{q_{n-1}}{\omega q_{n-1} + q_{n-2}}$ and $\omega = \frac{1}{\nu} - \frac{q_{n-2}}{q_{n-1}} > 1$. Next we consider the finite or infinite continued fraction of ω,

$$\omega = [b_n, b_{n+1}, b_{n+2}, \ldots].$$

Since $\omega > 1$, we conclude $b_j \in \mathbb{N}$, $j = n, n+1, n+2, \ldots$. Consequently,

$$\alpha = [b_0, b_1, \ldots, b_{n-1}, b_n, b_{n+1}, \ldots]$$

which is the continued fraction expansion for α and

$$\frac{p}{q} = \frac{p_{n-1}}{q_{n-1}} = [b_0, b_1, \ldots, b_{n-1}]$$

is indeed a convergent to α.

Lemma 1.3.4 *Assume the continued fraction expansion for α is given by*

$$\alpha = [a_0, a_1, \ldots, a_N, 1, 1, \ldots]. \tag{1.8}$$

Then

$$\lim_{n \to \infty} q_n^2 \left| \alpha - \frac{p_n}{q_n} \right| = \frac{1}{\sqrt{5}}.$$

A proof can be found in [45].

Theorem 1.3.5 (Hurwitz, [31]) *Let α be an irrational number.*

(i) *Then there are infinitely many rational numbers $\frac{p}{q}$ such that*

$$\left| \alpha - \frac{p}{q} \right| < \frac{1}{\sqrt{5}q^2}.$$

(ii) *If $\sqrt{5}$ is replaced by $C > \sqrt{5}$, then there are irrational numbers α for which statement (i) does not hold.*

Proof Claim (i) follows directly from Theorem 1.3.2, while claim (ii) follows from Theorem 1.3.3 and Lemma 1.3.4. Namely, if α is irrational and of the form (1.8), then according to Theorem 1.3.3, all solutions of $|\alpha - \frac{p}{q}| < \frac{1}{Cq^2}$ with $C > \sqrt{5}$ can be found among the convergents to α, however, in view of Lemma 1.3.4 this inequality is satisfied by only finitely many convergents to α.

Definition 1.3.6 An irrational number α is said to be a *quadratic irrational number* if it is the solution of some quadratic equation with rational coefficients.

Definition 1.3.7 The continued fraction expansion $[a_0, a_1, \ldots]$ of a real number α is said to be *eventually periodic* if there exist integers $m \geq 0$ and $h > 0$ such that

$$a_n = a_{n+h}, \quad \text{for all } n \geq m.$$

In this case the continued fraction expansion is denoted by

$$[a_0, a_1, \ldots, a_{m-1}, \overline{a_m, a_{m+1}, \ldots, a_{m+h-1}}].$$

The continued fraction expansion is said to be *periodic* if it is eventually periodic with $m = 0$.

Euler[6] and Lagrange[7] have proved an important characterization of quadratic irrationals in terms of their continued fraction expansion:

Theorem 1.3.8 (Euler 1737; Lagrange 1770) *A real number α is quadratic irrational if and only if its continued fraction expansion is eventually periodic.*

A proof can be found in [46].

Theorem 1.3.9 *Let $d > 1$ be an integer which is not a perfect square. Then the continued fraction expansion of $\sqrt{d}$ is of the form*

$$[a_0, \overline{a_1, \ldots, a_{n-1}, 2a_0}]$$

with $a_0 = \lfloor \sqrt{d} \rfloor, a_1 = a_{n-1}, a_2 = a_{n-2}, \ldots$.

A proof can be found in [41].

Remark 1.3.10 Let $E > 1$ be an integer which is not a perfect square. Let

$$\alpha_0 = \frac{s_0 + \sqrt{E}}{t_0}$$

be a quadratic irrational with $s_0, t_0 \in \mathbb{Z}$, $t_0 \neq 0$ such that $t_0 \mid (E - s_0^2)$. Then the partial quotients a_i are given by the recursion

$$a_i = \lfloor \sqrt{\alpha_i} \rfloor, \quad s_{i+1} = a_i t_i - s_i, \quad t_{i+1} = \frac{E - s_{i+1}^2}{t_i}, \quad \alpha_{i+1} = \left\lfloor \frac{s_{i+1} + \sqrt{E}}{t_{i+1}} \right\rfloor.$$

---

[6]Leonhard Euler (1707–1783), a Swiss mathematician.
[7]Joseph-Louis Lagrange (1736–1813), an Italian-French mathematician.

Example 1.3.11 For $d \in \mathbb{N}$,

$$\sqrt{2d(2d-1)} = [2d-1; \overline{2, 4d-2}],$$

$$\sqrt{2d(8d-1)} = [4d-1; \overline{1, 2, 1, 8d-2}].$$

Definition 1.3.12 An irrational number α is said to be *badly approximable* if there is a real number $c(\alpha) > 0$, depending only on α, such that

$$\left| \alpha - \frac{p}{q} \right| > \frac{c(\alpha)}{q^2}$$

for all rational numbers $\frac{p}{q}$.

Proposition 1.3.13 *If an irrational number α is badly approximable, then for every $\varepsilon > 0$, there are only finitely many rational numbers $\frac{p}{q}$ satisfying*

$$\left| \alpha - \frac{p}{q} \right| \leq \frac{1}{q^{2+\varepsilon}}.$$

Proof Let $c(\alpha)$ be a positive real number such that

$$\left| \alpha - \frac{p}{q} \right| > \frac{c(\alpha)}{q^2}$$

for all rational numbers $\frac{p}{q}$. Now suppose that there are infinitely many rational numbers $\frac{p}{q}$ satisfying

$$\left| \alpha - \frac{p}{q} \right| \leq \frac{1}{q^{2+\varepsilon}}.$$

Consequently, there exists a rational number $\frac{p_1}{q_1}$ for which the inequality

$$q_1^\varepsilon > \frac{1}{c(\alpha)}$$

holds. It follows that

$$\left| \alpha - \frac{p_1}{q_1} \right| \leq \frac{1}{q_1^2 q_1^\varepsilon} < \frac{c(\alpha)}{q_1^2},$$

which is a contradiction.

Corollary 1.3.14 *For quadratic irrational numbers α and any $\varepsilon > 0$, there are only finitely many rational numbers $\frac{p}{q}$ satisfying*

$$\left| \alpha - \frac{p}{q} \right| \leq \frac{1}{q^{2+\varepsilon}}.$$

A proof can be found in [44]. Hence, the exponent in the upper bound $\frac{1}{q^2}$ in Dirichlet's Theorem 1.1.1 cannot be improved for real quadratic irrational numbers.

More on Diophantine approximation can be found in [13, 19, 20, 44, 45].

1.4 Algebraic and Transcendental Numbers

1.4.1 Basic Theorems and Definitions

Definition 1.4.1 Given fields F and E such that $F \subseteq E$, then E is called an *extension* of F, denoted by $F \leq E$ or E/F. The dimension of the extension is called the *degree* of the extension and it is abbreviated by $[E : F]$. If $[E : F] = n < \infty$, E is said to be a *finite* extension of F.

Remark 1.4.2 It can be shown that, if f is a nonconstant polynomial over the field F, then there is an extension E of the field F and $\alpha \in E$ such that

$$f(\alpha) = 0.$$

Hence, the field extension of the initial field F is often obtained from the field F by adjoining a root α of a nonconstant polynomial f over the field F; in this case the minimal field containing α and F is denoted by $F(\alpha)$.

Definition 1.4.3 A real number α is called *algebraic* over $\mathbb{Q}$ if it is a root of a polynomial equation with coefficients in $\mathbb{Q}$. A real number is said to be *transcendental* if it is not algebraic.

Definition 1.4.4 The *minimal polynomial* p of an algebraic number α over $\mathbb{Q}$ is the uniquely determined irreducible monic polynomial of minimal degree with rational coefficients satisfying

$$p(\alpha) = 0.$$

Elements that are algebraic over $\mathbb{Q}$ and have the same minimal polynomial are called *conjugates* over $\mathbb{Q}$.

Definition 1.4.5 Let α be an algebraic number. Then the degree of α is the degree of the minimal polynomial of α over $\mathbb{Q}$.

Remark 1.4.6 If

$$X^d + r_{d-1}X^{d-1} + \cdots + r_1 X + r_0$$

is the minimal polynomial of an algebraic number α over $\mathbb{Q}$, then multiplication by the least common multiple of the denominators of the coefficients r_i, $i = 0, \ldots,$

$d - 1$, produces a unique polynomial P with $P(\alpha) = 0$ having the form

$$P(X) = a_d X^d + a_{d-1} X^{d-1} + \cdots + a_1 X + a_0,$$

where the coefficients a_i, $i = 0, \ldots, d$, are relatively prime integers and $a_d > 0$. We will call this polynomial the *minimal polynomial of* α over $\mathbb{Z}$.

Example 1.4.7 A real number is rational if and only if it is an algebraic number of degree 1.

Example 1.4.8 We shall show that

$$x = \frac{10^{\frac{2}{3}} - 1}{\sqrt{-3}}$$

is an algebraic number.

To see that, we write $\sqrt{-3}x + 1 = 10^{2/3}$ and get $(\sqrt{-3}x + 1)^3 = 100$. Expanding the left-hand side, it follows that

$$3\sqrt{-3}x^3 + 9x^2 - 3\sqrt{-3}x + 99 = 0.$$

Dividing by $3\sqrt{-3}$, we get

$$x^3 - \sqrt{-3}x^2 - x - 11\sqrt{-3} = 0.$$

From this we deduce the minimal polynomial of x as

$$P(x) = \left((x^3 - x) - \sqrt{-3}(x^2 + 11) \right)\left((x^3 - x) + \sqrt{-3}(x^2 + 11) \right) =$$
$$= x^6 + x^4 + 67x^2 + 363.$$

1.4.2 Liouville's Theorem

The main task in Diophantine approximation is to figure out how well a real number α can be approximated by rational numbers. In view of this problem, as we have mentioned earlier, a rational number $\frac{p}{q}$ is considered to be a "good" approximation of a real number α if the absolute value of the difference between $\frac{p}{q}$ and α may not decrease when $\frac{p}{q}$ is replaced by another rational number with a *smaller* denominator. This problem was solved during the 18th century by means of continued fractions.

Knowing the "best" approximation of a given number, the main problem is to find sharp upper and lower bounds of the mentioned difference, expressed as a function of the denominator.

It appears that these bounds depend on the nature of the real numbers to be approximated: the lower bound for the approximation of a rational number by another rational number is larger than the lower bound for algebraic numbers, which is itself larger than the lower bound for all real numbers. Thus, a real number that may be better approximated than an algebraic number is certainly a *transcendental number*. This statement had been proved by Liouville[8] in 1844, and it produced the first explicit transcendental numbers. The later proofs on the transcendency of π and e were obtained by a similar idea.

Theorem 1.4.9 (Liouville 1844) *Let α be a real algebraic number of degree $d \geq 2$. Then there exists a constant $c(\alpha) > 0$, depending only on α, such that*

$$\left| \alpha - \frac{p}{q} \right| \geq \frac{c(\alpha)}{q^d}$$

for all rational numbers $\frac{p}{q}$.

There are a few ways of proving Liouville's Theorem 1.4.9: one can be found in [12]; a second, and maybe the most familiar one, uses the mean value theorem, an interesting variant concerning the constant $c(\alpha)$ is given in [45].

Corollary 1.4.10 *Let α be a real algebraic number of degree $d \geq 2$. For every $\delta > 0$, there are only finitely many rational numbers $\frac{p}{q}$ satisfying the inequality*

$$\left| \alpha - \frac{p}{q} \right| \leq \frac{1}{q^{d+\delta}}.$$

A proof can be found in [44].

We observe that the denominators of rational numbers which are getting closer and closer to a fixed real number α grow *arbitrarily large*. Liouville's Theorem 1.4.9 proves that this is indeed the case for real algebraic numbers α. For example, Liouville's Theorem 1.4.9 implies that, if $\frac{p}{q}$ is within a distance $\frac{1}{10^{10}}$ of α, then

$$q^d \geq 10^{10} c(\alpha).$$

Remark 1.4.11 Consider the rational numbers $\frac{p}{q}$ and $\frac{p'}{q'}$ satisfying

$$q \left| \alpha - \frac{p}{q} \right| < q' \left| \alpha - \frac{p'}{q'} \right|.$$

The theory of continued fractions provides an efficient algorithm for constructing the best approximations to α in this case.

Liouville's Theorem 1.4.9 can be used to find transcendental numbers explicitly. His construction of transcendental numbers (1844) predates Cantor's[9] proof (1874) of their existence.

---

[8] Joseph Liouville (1809–1882), a French mathematician.
[9] Georg Ferdinand Ludwig Philipp Cantor (1845–1918), a German mathematician.

Example 1.4.12 Let

$$\alpha = \sum_{n=1}^{\infty} \frac{1}{10^{n!}} = 0.110001000000000000000000001000....$$

The digit 1 appears in the 1st, 2nd, 6th, 24th, ..., $(n!)$th,... decimal place. For $k \geq 1$, let $q(k) = 10^{k!}$ and $p(k) = 10^{k!} \sum_{n=1}^{k} \frac{1}{10^{n!}}$. We notice that $p(k)$ and $q(k)$ are relatively prime integers, $\frac{p_k}{q_k} = \sum_{n=1}^{k} \frac{1}{10^{n!}}$ and

$$\left| \alpha - \frac{p(k)}{q(k)} \right| = \sum_{n=k+1}^{\infty} \frac{1}{10^{n!}}.$$

Comparing with a geometric series, we find

$$\sum_{n=k+1}^{\infty} \frac{1}{10^{n!}} < \frac{1}{10^{(k+1)!}} \sum_{n=0}^{\infty} \frac{1}{10^n} = \frac{10}{9} \frac{1}{q(k)^{k+1}},$$

and

$$\left| \alpha - \frac{p(k)}{q(k)} \right| < \frac{\frac{10}{9}}{q(k)^{k+1}}.$$

Finally, we observe that α does not satisfy Liouville's Theorem 1.4.9. It follows from the calculations above that for any $c > 0$ and any $d > 0$, selecting k such that $\frac{10}{9} < c \cdot q(k)^{k+1-d}$, leads to

$$\left| \alpha - \frac{p(k)}{q(k)} \right| < \frac{c}{q(k)^d}$$

for large k. Thus, α must be a transcendental number.

In 1873, Charles Hermite[10] proved that e is transcendental and nine years later Ferdinand von Lindemann[11] proved the transcendence of π. Hermite even showed that e^a is transcendental when a is algebraic and nonzero. This approach was generalized by Weierstrass[12] to the Lindemann-Weierstrass theorem.

Theorem 1.4.13 (Hermite) *e is transcendental.*

A proof can be found in [4, 28].

---

[10]Charles Hermite (1822–1901), a French mathematician.

[11]Ferdinand von Lindemann (1852–1939), a German mathematician.

[12]Karl Weierstrass (1815–1897), a German mathematician.

Theorem 1.4.14 (Lindemann) π *is transcendental.*

A proof can be found in [34].

Theorem 1.4.15 (Lindemann, Weierstrass) *Let* $\beta_1, \ldots, \beta_n$ *be nonzero algebraic numbers and* $\alpha_1, \ldots, \alpha_n$ *distinct algebraic numbers. Then*

$$\beta_1 e^{\alpha_1} + \cdots + \beta_n e^{\alpha_n} \neq 0.$$

A proof can be found in [4].

Corollary 1.4.16 *If* α *is a nonzero algebraic integer, then*

$$e^{\alpha}, \ \sin \alpha, \ \cos \alpha$$

are transcendental numbers.

Definition 1.4.17 A real number α is a *Liouville number* if for every positive integer n, there exist integers p, q with $q > 1$ such that

$$0 < \left| \alpha - \frac{p}{q} \right| < \frac{1}{q^n}.$$

Remark 1.4.18 It is known that π and e are not Liouville numbers (see [37, 41], respectively). Mahler [38] found conditions on the expansion of a real number α in base g that imply that α is transcendental but not a Liouville number. One such example is the decimal number called Champernowne's constant (Mahler's number).

Theorem 1.4.19 *All Liouville numbers are transcendental.*

A proof can be found in [46].

Example 1.4.20 We show that

$$\alpha = \sum_{j=0}^{\infty} \frac{1}{2^{j!}}$$

is a Liouville number.

First, we observe that the binary expansion of α has arbitrarily long strings of 0's, so it cannot be rational. Fix a positive integer n and consider $\frac{p}{q} = \sum_{j=0}^{n} \frac{1}{2^{j!}}$ with p and $q = 2^{j!} > 1$ integers. Then

$$0 < \left| \alpha - \frac{p}{q} \right| = \sum_{j=n+1}^{\infty} \frac{1}{2^{j!}} < \sum_{j=(n+1)!}^{\infty} \frac{1}{2^{j!}} = \frac{1}{2^{(n+1)!-1}} \leq \frac{1}{2^{n(n!)}} = \frac{1}{q^n},$$

which proves that α is indeed a Liouville number.

1.4.3   Roth's Theorem

It is natural to ask for stronger versions of Liouville's Theorem 1.4.9. First improvements were made by Thue,[13] Siegel[14] and Dyson.[15] In 1955, K.F. Roth[16] proved the most far-reaching extension, now known as the Thue-Siegel-Roth theorem, but also just as Roth's theorem, for which he was awarded a Fields Medal in 1958. We quote from Roth's paper [43].

Theorem 1.4.21 (Roth 1955) *Let α be a real algebraic number of degree $d \geq 2$. Then, for every $\delta > 0$, the inequality*

$$\left| \alpha - \frac{p}{q} \right| \leq \frac{1}{q^{2+\delta}}$$

has only finitely many rational solutions $\frac{p}{q}$.

A complete proof of Roth's Theorem 1.4.21 can be found in [45], and for a generalized version we refer to [30].

Corollary 1.4.22 *Let α be an algebraic number of degree $d \geq 2$. Then, for every $\delta > 0$, there is a constant $c(\alpha, \delta) > 0$ such that*

$$\left| \alpha - \frac{p}{q} \right| \geq \frac{c(\alpha, \delta)}{q^{2+\delta}}$$

for all rational numbers $\frac{p}{q}$.

A proof can be found in [44]. Note that, if α is a real algebraic number of degree 2, then Liouville's Theorem 1.4.9 is stronger than Roth's Theorem 1.4.21. Whereas the latter one gives the estimate

$$\left| \alpha - \frac{p}{q} \right| \geq \frac{c(\alpha, \delta)}{q^{2+\delta}}$$

for every $\delta > 0$, Liouville's Theorem 1.4.9 states that there is a constant $c(\alpha) > 0$ such that

$$\left| \alpha - \frac{p}{q} \right| \geq \frac{c(\alpha)}{q^2}$$

for all rational numbers $\frac{p}{q}$. The analogous result for real algebraic numbers α of degree ≥ 3, namely, that there is a constant $c(\alpha) > 0$ such that an inequality of the form

---

[13] Axel Thue (1863–1922), a Norwegian mathematician.

[14] Carl Ludwig Siegel (1896–1981), a German mathematician.

[15] Freeman Dyson (1923), an English-born American mathematician.

[16] Klaus Friedrich Roth (1925–2015), a German-born British mathematician.

$$\left| \alpha - \frac{p}{q} \right| \geq \frac{c(\alpha)}{q^2}$$

holds for all rational numbers $\frac{p}{q}$, is conjectured to be false for all such α. However, this is not known at the present time to be false for a single real algebraic number.

Champernowne's Constant

The transcendental numbers (except e and π) we have encountered so far are transcendental because their decimal (or dyadic) expansions have infinitely many runs of zeros whose lengths grow so quickly that the simple truncation of the decimal (or dyadic) expansion before each run of zeros leads to amazingly good approximations. However, for most numbers a collection of best rational approximations is not so easily detected from their decimal expansions. For most numbers even the best rational approximations are not close enough to allow us to conclude transcendency.

As an illustration of the difficulty of finding suitable rational approximations in general, we consider Champernowne's constant (resp. Mahler's[17] number).

Definition 1.4.23 *Champernowne's constant (Mahler's number)*

$$\mathcal{M} = 0.12345678910111213141516171819 2021\ldots$$

is[18] the number obtained by concatenating the positive integers in base 10 and interpreting them as decimal digits to the right of a the decimal point.

We mention a method worked out in detail in [11]. Firstly, rational approximations created by long runs of zeros are used, and $\mathcal{M}$ is truncated just after the 1 that appears whenever a power of 10 is reached. Following this truncation procedure, we see that the number of decimal digits before each run of zeros exceeds the length of that run by far. For example, to get a run of just one zero, $\mathcal{M}$ has to be truncated after 10 digits, i.e., 0.123456789. In general, if we want to come across a run of k zeros, we have to travel on the order of $k \cdot 10^k$ digits from the previous run of $k - 1$ zeros. Thus, this truncation method cannot generate rational approximations having relatively small denominators that are sufficiently close to $\mathcal{M}$ in order to prove transcendence via Liouville's Theorem 1.4.9.

Using a more clever construction to build rational approximations to $\mathcal{M}$, described in [11], one can find approximations all having relatively small denominators that allow to apply Liouville's Theorem 1.4.9 to derive the following partial result.

Theorem 1.4.24 (Mahler 1937) *The number*

$$\mathcal{M} = 0.12345678910111213141516171819 2021\ldots$$

is either a transcendental number or an algebraic number of a degree at least 5.

---

[17]Kurt Mahler (1903–1988), a German/British mathematician.

[18]OEIS A033307.

A proof can be found in [11, 38]. Theorem 1.4.24 does not guarantee that $\mathcal{M}$ is transcendental, however, it does imply that $\mathcal{M}$ is neither a quadratic irrational, nor a cubic or even quartic algebraic number. A more advanced analysis building on Liouville's Theorem 1.4.9 and Roth's Theorem 1.4.21 would ensure that $\mathcal{M}$ is transcendental.

Theorem 1.4.25 (Mahler 1937) *Champernowne's constant $\mathcal{M}$ is a transcendental number.*

A proof can be found in [11, 38].

1.4.4   Thue's Theorem

Thue's work was already a major breakthrough for those kind of questions:

Theorem 1.4.26 (Thue) *Let α be a real algebraic number of degree d. Then, for every $\delta > 0$, the inequality*

$$\left| \alpha - \frac{p}{q} \right| \leq \frac{1}{q^{\frac{1}{2}d+1+\delta}}$$

has only finitely many rational solutions $\frac{p}{q}$.

A proof can be found in [44].

Corollary 1.4.27 *Let α be a real algebraic number of degree d. Then, for every $\delta > 0$, there is a constant $c(\alpha, \delta) > 0$ such that*

$$\left| \alpha - \frac{p}{q} \right| \geq \frac{c(\alpha, \delta)}{q^{\frac{1}{2}d+1+\delta}}$$

for all rational numbers $\frac{p}{q}$.

A proof can be found in [44].

The proof of Roth's Theorem 1.4.21 is immensely more complex than those for the theorems of Liouville and Thue, even though, the framework is in essence the same. The proof of Roth's Theorem 1.4.21 is not effective, that is, as noted in [30], for a given α, the proof does not provide a method that guarantees to find the finitely many rational numbers $\frac{p}{q}$ satisfying

$$\left| \alpha - \frac{p}{q} \right| \leq \frac{1}{q^{2+\delta}}.$$

In other words, the proof does not give a lower bound on $c(\alpha, \delta)$. It is with respect to the work of Thue and Siegel that Roth's Theorem 1.4.21 is often named the Thue-Siegel-Roth theorem.

We give an application of the Thue–Siegel–Roth theorem to Diophantine equations. It follows from the fact that the approximation exponent of an algebraic number α of degree $d \geq 3$ is strictly less than d. It appears that the full force of Roth's Theorem 1.4.21 is not needed, Thue's Theorem 1.4.26 is sufficient.

Theorem 1.4.28 *Let*

$$a_d X^d + a_{d-1} X^{d-1} + \cdots + a_1 X + a_0 \in \mathbb{Z}[X] \tag{1.9}$$

be an irreducible polynomial over $\mathbb{Q}$ of degree $d \geq 3$. Then, for every nonzero integer m, the Diophantine equation

$$a_d X^d + a_{d-1} X^{d-1} Y + \cdots + a_1 X Y^{d-1} + a_0 Y^d = m \tag{1.10}$$

has only finitely many integer solutions (p, q).

A proof can be found in [47]. An equation of the form (1.10) is called a *Thue equation*. It is interesting to note that equations of the form $X^2 - dY^2 = 1$, where d is a positive and a square free integer, the so called Pellian equations, have infinitely many integer solutions, but equations of the form $X^3 - dY^3 = 1$ with an arbitrary integer d have at most finitely many integer solutions. The books [11, 46] provide nice introductions to various aspects of Diophantine approximation, transcendence theory and Diophantine equations.

2   Linear Forms in Logarithms

2.1   *Introduction*

Hilbert's problems form a list of twenty-three major problems in mathematics collected, proposed and published by D. Hilbert[19] in 1900. The problems were all unsolved at the time and several of them turned out to be very influential for 20th century mathematics. Hilbert believed that new machinery and methods were needed for solving these problems. He presented ten of them at the International Congress of Mathematicians in Paris in 1900. The complete list of his 23 problems was published later, most notably an English translation appeared 1902 in the *Bulletin of the American Mathematical Society*.

Hilbert's seventh problem, entitled "irrationality and transcendence of certain numbers", is dealing with the transcendence of the number

$$\alpha^\beta$$

for algebraic $\alpha \neq 0, 1$ and irrational algebraic β. He believed that the proof of this problem would only be given in more distant future than proofs of Riemann's hypothesis or Fermat's last theorem. Even though Hilbert was mistaken, he was correct when

---

[19]David Hilbert (1862–1943), a German mathematician.

he expressed his belief that the proof of that problem would be extremely intriguing and influential for 20th century mathematics. The seventh problem was solved independently by Gelfond[20] and Schneider[21] in 1935. They proved that, if $\alpha_1, \alpha_2 \neq 0$ are algebraic numbers such that $\log \alpha_1, \log \alpha_2$ are linearly independent over $\mathbb{Q}$, then

$$\beta_1 \log \alpha_1 + \beta_2 \log \alpha_2 \neq 0$$

for all algebraic numbers β_1, β_2.

In 1935, Gelfond found a lower bound for the absolute value of the linear form

$$\Lambda = \beta_1 \log \alpha_1 + \beta_2 \log \alpha_2 \neq 0.$$

He proved that

$$\log |\Lambda| \gg -h(\Lambda)^{\kappa},$$

where $h(\Lambda)$ is the logarithmic height of the linear form Λ, $\kappa > 5$ and $\gg$ is Vinogradov's notation for an inequality that is valid up to an unspecified constant factor. Gelfond also noticed that generalization of his results would lead to a powerful new analytic method by which mathematicians could prove a huge amount of unsolved problems in number theory.

2.2 Basic Theorems and Definitions

In 1966 in 1967, A. Baker[22] gave in his papers "Linear forms in logarithms of algebraic numbers I, II, III", [1–3] an effective lower bound on the absolute value of a nonzero linear form in logarithms of algebraic numbers, that is, for a nonzero expression of the form

$$\sum_{i=1}^{n} b_i \log \alpha_i,$$

where $\alpha_1, \ldots, \alpha_n$ are algebraic numbers and $b_1, \ldots, b_n$ are integers. This result initiated the era of effective resolution of Diophantine equations that can be reduced to exponential ones (where the unknown variables are in the exponents). The generalization of the Gelfond-Schneider theorem was only the beginning of a new and very interesting branch in number theory called *Baker's theory*.

Definition 2.2.1 Let $\alpha_1, \alpha_2, \ldots, \alpha_n$ be n (real or complex) numbers. We call $\alpha_1, \alpha_2, \ldots, \alpha_n$ *linearly dependent* over the rationals (equivalently integers) if there are rational numbers (integer numbers) $r_1, r_2, \ldots, r_n$, not all zero, such that

---

[20] Alexander Osipovich Gelfond (1906–1968), a Soviet mathematician.

[21] Theodor Schneider (1911–1988), a German mathematician.

[22] Alan Baker (1939), an English mathematician.

$$r_1\alpha_1 + r_2\alpha_2 + \cdots + r_n\alpha_n = 0.$$

If $\alpha_1, \alpha_2, \ldots, \alpha_n$ are not linearly dependent over the rationals (integers), they are *linearly independent* over the rationals (integers).

Definition 2.2.2 A *linear form in logarithms* of algebraic numbers is an expression of the form

$$\Lambda = \beta_0 + \beta_1 \log \alpha_1 + \beta_2 \log \alpha_2 + \cdots + \beta_n \log \alpha_n,$$

where α_i, $i = 1, \ldots, n$ and β_i, $i = 0, \ldots, n$ are complex algebraic numbers and log denotes any determination of the logarithm.

Remark 2.2.3 We are interested in the degenerate case when $\beta_0 = 0$ and $\beta_i \in \mathbb{Z}$, $i = 1, \ldots, n$. In the sequel we write $\beta_i = b_i$, $i = 1, \ldots, n$ and log always represents the principal value of the complex logarithm.

A generalization of the Gelfond-Schneider theorem to arbitrarily many logarithms was obtained by Baker in 1966 [1]. In 1970, he was awarded the Fields Medal for his work in number theory, especially in the areas of transcendence and Diophantine geometry. One of his major contributions is the following

Theorem 2.2.4 (Baker 1966) *If $\alpha_1, \alpha_2, \ldots, \alpha_n \neq 0, 1$ are algebraic numbers such that $\log \alpha_1, \log \alpha_2, \ldots, \log \alpha_n, 2\pi i$ are linearly independent over the rationals, then*

$$\beta_0 + \beta_1 \log \alpha_1 + \cdots + \beta_n \log \alpha_n \neq 0$$

for any algebraic numbers $\beta_0, \beta_1, \ldots, \beta_n$ that are not all zero.

A proof can be found in [1].

Remark 2.2.5 Linear independence over the rationals implies linear independence over the algebraic numbers (see [4]).

Theorem 2.2.6 (Baker 1967) *The number $e^{\beta_0}\alpha_1^{\beta_1} \ldots \alpha_n^{\beta_n}$ is transcendental for all nonzero algebraic numbers α_i, $i = 1, \ldots, n$ and β_i, $i = 0, \ldots, n$. Furthermore, the number $\alpha_1^{\beta_1} \ldots \alpha_n^{\beta_n}$ is transcendental if $1, \beta_1, \ldots, \beta_n$ are linearly independent over the rationals.*

A proof can be found in [3].

Definition 2.2.7 The *height H* of a rational number $\frac{p}{q}$ is defined by

$$H\left(\frac{p}{q}\right) = \max\{|p|, |q|\}.$$

Definition 2.2.8 Let $\mathbb{L}$ be a number field of the degree D, $\alpha \in \mathbb{L}$ an algebraic number of degree $d \mid D$ and let $\sum_{0 \leq k \leq d} a_k X^k$ be its minimal polynomial in $\mathbb{Z}[X]$ with $a_d \neq 0$. We define the *absolute logarithmic height* $h(\alpha)$ by

$$h(\alpha) = \frac{1}{d}\left(\log(|a_d|) + \sum_{1 \le i \le d} \max\{\log(|\alpha_i|, 0)\}\right),\qquad(2.1)$$

where α_i are the conjugates of α.

Example 2.2.9 The absolute logarithmic height h of a rational number $\frac{p}{q}$ is

$$h\left(\frac{p}{q}\right) = \log\max\{|p|, |q|\}.$$

Example 2.2.10 Let $\alpha = \sqrt{2}$. The absolute logarithmic height of α is

$$h(\alpha) = h(\sqrt{2}) = \frac{1}{2}\left(\log|\sqrt{2}| + \log|-\sqrt{2}|\right) = \frac{1}{2}\log 2.$$

Example 2.2.11 Let

$$\alpha_1 = \frac{1}{\sqrt{3}+\sqrt{5}}.$$

Before calculating the absolute logarithmic height of α_1, we compute the degree of the field extension $\mathbb{Q}\left(\frac{1}{\sqrt{3}+\sqrt{5}}\right)$ over $\mathbb{Q}$ as

$$\left[\mathbb{Q}\left(\frac{1}{\sqrt{3}+\sqrt{5}}\right):\mathbb{Q}\right] = 4,$$

so the degree of the algebraic number $\alpha_1 = \frac{1}{\sqrt{3}+\sqrt{5}}$ over $\mathbb{Q}$ is 4.
The minimal polynomial of α_1 is given by

$$P_{\alpha_1}(x) = \left(x - \frac{1}{\sqrt{3}+\sqrt{5}}\right)\left(x - \frac{1}{-\sqrt{3}-\sqrt{5}}\right) \times$$
$$\times \left(x - \frac{1}{-\sqrt{3}+\sqrt{5}}\right)\left(x - \frac{1}{\sqrt{3}-\sqrt{5}}\right)$$
$$= x^4 - 4x^2 + \frac{1}{4},$$

over the rationals, and

$$P_{\alpha_1}(x) = 4x^4 - 16x^2 + 1$$

over the integers. The conjugates of α_1 are

$$\alpha_1 = \frac{1}{\sqrt{3}+\sqrt{5}}, \ \alpha_2 = \frac{1}{-\sqrt{3}+\sqrt{5}}, \ \alpha_3 = \frac{1}{\sqrt{3}-\sqrt{5}}, \ \alpha_4 = \frac{1}{-\sqrt{3}-\sqrt{5}}.$$

Hence, using (2.1), the absolute logarithmic height of α_1 is equal to

$$h(\alpha_1) = h\left(\frac{1}{\sqrt{3}+\sqrt{5}}\right)$$

$$= \frac{1}{4}\left(\log|4| + \max\left\{\log\left|\frac{1}{\sqrt{3}+\sqrt{5}}\right|, 0\right\} + \max\left\{\log\left|\frac{1}{-\sqrt{3}+\sqrt{5}}\right|, 0\right\} +\right.$$

$$\left. + \max\left\{\log\left|\frac{1}{\sqrt{3}-\sqrt{5}}\right|, 0\right\} + \max\left\{\log\left|\frac{1}{-\sqrt{3}-\sqrt{5}}\right|, 0\right\}\right)$$

$$= 0.689146.$$

Let $\mathbb{L}$ be a number field of degree D, $\alpha_1, \alpha_2, \ldots, \alpha_n$ nonzero elements of $\mathbb{L}$ and let $b_1, b_2, \ldots, b_n$ be integers. We define

$$B = \max\{|b_1|, |b_2|, \ldots, |b_n|\}$$

and

$$\Lambda^* = \alpha_1^{b_1}\alpha_2^{b_2}\ldots\alpha_n^{b_n} - 1.$$

We wish to bound $|\Lambda^*|$ from below, assuming that it is nonzero.

Since $\log(1+x)$ is asymptotically equal to x as $|x|$ tends to 0, our problem consists of finding a lower bound for the linear form in logarithms

$$\Lambda = b_1 \log \alpha_1 + \cdots + b_n \log \alpha_n + b_{n+1}\log(-1),$$

where $b_{n+1} = 0$ if $\mathbb{L}$ is real and $|b_{n+1}| \le nB$, otherwise.

Definition 2.2.12 Let $A_1, A_2, \ldots, A_n$ be real numbers such that

$$A_j \ge h'(\alpha_j) := \max\{Dh(\alpha_j), |\log\alpha_j|, 0.16\}, \quad 1 \le j \le n.$$

Then h' is called the *modified height* with respect to the field $\mathbb{L}$.

A. Baker, E.M. Matveev[23] and G. Wüstholz[24] proved the following theorems.

Theorem 2.2.13 (Baker–Wüstholz 1993) *Assume that*

$$\Lambda = b_1 \log \alpha_1 + \cdots + b_n \log \alpha_n \ne 0$$

for algebraic α_i and integers b_i, $i = 1, \ldots, n$. Then

$$\log|\Lambda| \ge -18(n+1)!n^{n+1}(32D)^{n+2}\log(2nD)h''(\alpha_1)\ldots h''(\alpha_n)\log B,$$

where D is the degree of the extension $\mathbb{Q}(\alpha_1, \ldots, \alpha_n)$, $B = \max\{|b_i|, i = 1, \ldots, n\}$ and $h''(\alpha) = \max\{h(\alpha), \frac{1}{D}|\log(\alpha)|, \frac{1}{D}\}$.

---

[23]Eugene Mikhailovich Mateveev (1955), a Russian mathematician.

[24]Gisbert Wüstholz (1948), a German mathematician.

A proof can be found in [6].

Theorem 2.2.14 (Matveev 2001) *Assume that Λ^* is nonzero. Then*

$$|\Lambda^*| > -3 \cdot 30^{n+4}(n+1)^{5.5}D^2 A_1 \ldots A_n(1 + \log D)(1 + \log nB).$$

If $\mathbb{L}$ is real, then

$$\log|\Lambda^*| > -1.4 \cdot 30^{n+3}(n+1)^{4.5}D^2 A_1 \ldots A_n(1 + \log D)(1 + \log B).$$

Theorem 2.2.15 (Matveev 2001) *Assume that*

$$\Lambda = b_1 \log \alpha_1 + \cdots + b_n \log \alpha_n \neq 0$$

for algebraic α_i and integers b_i, $i = 1, \ldots, n$. Then

$$\log|\Lambda| > -2 \cdot 30^{n+4}(n+1)^6 D^2 A_1 \ldots A_n(1 + \log D)(1 + \log B),$$

where $B = \max\{|b_i|,\ 1 \le i \le n\}$.

Proofs for Matveev's theorems can be found in [39].

2.3   A Variation of Baker–Davenport Lemma

For a real number x we introduce the notation

$$||x|| = \min\{|x - n| : n \in \mathbb{Z}\}$$

for the distance from x to the nearest integer.

The following result is a variation of a lemma of Baker and Davenport[25] [5], it is due to Dujella[26] and Pethő,[27] [22].

Lemma 2.3.1 *Let N be a positive integer, $\frac{p}{q}$ a convergent of the continued fraction expansion of an irrational number κ such that $q > 6N$ and let μ be some real number. Let $\varepsilon = ||\mu q|| - N||\kappa q||$. If $\varepsilon > 0$, then there is no solution to the inequality*

$$0 < m\kappa - n + \mu < AB^{-m}$$

in positive integers m and n with

$$\frac{\log(Aq/\varepsilon)}{\log B} \leq m \leq N.$$

Proof Suppose that $0 \leq m \leq N$. Then

$$m(\kappa q - p) + mp - nq + \mu q < qAB^{-m}.$$

Thus,

$$qAB^{-m} > |\mu q - (nq - mp)| - m||\kappa q|| \geq ||\mu q|| - N||\kappa q|| := \varepsilon,$$

from where we deduce that

$$m < \frac{\log(Aq/\varepsilon)}{\log B}.$$

Remark 2.3.2 The method from Lemma 2.3.1 is called Baker–Davenport reduction.

Example 2.3.3 Find all nonnegative integers that satisfy

$$0 < |x_1 \log 2 - x_2 \log 3 + \log 5| < 40e^{-X}, \tag{2.2}$$

where $X = \max\{x_1, x_2\} \leq 10^{30}$.
    Let

$$x_1 \log 2 - x_2 \log 3 + \log 5 > 0.$$

First, we divide (2.2) by $\log 3$, in order to get inequalities of the form as in Lemma 2.3.1. We get

$$0 < x_1 \frac{\log 2}{\log 3} - x_2 + \frac{\log 5}{\log 3} < \frac{40}{\log 3} e^{-X}.$$

Now, we define

$$\kappa = \frac{\log 2}{\log 3}, \quad \mu = \frac{\log 5}{\log 3}, \quad A = \frac{40}{\log 3}, \quad B = e.$$

We observe that the inequalities $A > 0$, $B > 1$ are satisfied.
    We shall try to find a convergent $\frac{p}{q}$ of the continued fraction expansion of κ that satisfies the condition $q > 6N$. Since κ does not have a finite or periodic continued fraction expansion, we give only the first 25 terms of its continued fraction expansion:

$$\kappa = [0, 1, 1, 1, 2, 2, 3, 1, 5, 2, 23, 2, 2, 1, 1, 55, 1, 4, 3, 1, 1, 15, 1, 9, 2, \dots].$$

The first convergent $\frac{p}{q}$ that satisfies the inequality $q > 6N$ is

$$\frac{p}{q} = \frac{35270892459770675836042178475339}{55903041915705101922536695520222}.$$

We therefore obtain

$$||\kappa q||N \approx 0.007651391, \qquad ||\mu q|| \approx 0.466714899.$$

Hence,

$$\varepsilon = ||\mu q|| - ||\kappa q||N = 0.4590635078 > 0.$$

The given inequality does not have any solution in integers m such that

$$\frac{\ln\left(\frac{40}{\log 3} \cdot q \cdot \frac{1}{0.4590635}\right)}{\ln e} \leq m \leq N.$$

We observe that $m \leq 77.4746$, so $N = 77$. Repeating Baker–Davenport reduction one more time, but now with $N = 77$, we find convergents $\frac{p}{q}$ of the continued fraction expansion of κ that satisfies the condition $> 6N$, $N = 77$. The first such convergent is

$$\frac{p}{q} = \frac{306}{485}.$$

We get

$$||\kappa q||N \approx 0.071647126, \qquad ||\mu q|| \approx 0.487842451.$$

Finally, we obtain

$$\varepsilon \approx 0.4161953257.$$

After applying Baker–Davenport reduction, we get $m \leq 10.6556$, resp. $N = 10$. Therefore, we can find pairs (x_1, x_2) satisfying the introduced inequalities. Using a simple computer algorithm, it turns out that the following pairs

$$(0, 0), (0, 1), (1, 0), (1, 1), (1, 2), (2, 0), (2, 1), (2, 2), (3, 0), (3, 1), (3, 2), (3, 3), (4, 3)$$

satisfy (2.2).
    For

$$x_1 \log 2 - x_2 \log 3 + \log 5 < 0,$$

we have

$$0 < x_2 \frac{\log 3}{\log 2} - x_1 - \frac{\log 5}{\log 2} < \frac{40}{\log 2} e^{-x}.$$

We define

$$\kappa = \frac{\log 3}{\log 2}, \quad \mu = -\frac{\log 5}{\log 2}, \quad A = \frac{40}{\log 2}, \quad B = e.$$

Then the inequalities $A > 0$, $B > 1$ are satisfied. The next step is to find a convergent $\frac{p}{q}$ of the continued fraction expansion of κ satisfying $q > 6N$. For

$$\frac{p}{q} = \frac{15220760977433336077811000045449522888}{960323097266207036440783078900790949}$$

we get

$$\varepsilon = 0.124406274 > 0.$$

Analogously, we find $N = 89$ after the first reduction and $N = 13$ after the second one. Finally, we get that the following pairs

$$(0, 2), (0, 3), (1, 3), (2, 3), (2, 4), (3, 4), (4, 4), (5, 5).$$

that satisfy (2.2), as well.

2.4 Applications

For the applications of linear forms in logarithms to Diophantine equations, the strategy is as follows: first, we use various algebraic manipulations to associate "relatively big" solutions of the equations to a "very small" value of the specific linear form in logarithms which implies that we are able to find upper bound for values of the linear form in the logarithms that corresponds to a solution of the equation. If we compare that upper bound with the lower bound (using the Baker–Wüstholz Theorem 2.2.13 or Matveev's Theorems 2.2.14 and 2.2.15), we get an absolute upper bound M for the absolute values of the unknowns of the equations.

It often happens that the upper bound M is not too large and using various methods, including reductions and sieves, we can find the complete set of solutions below M. In order to realize this, it is crucial to get a reasonably small value for M. Its size is directly related to the size of the "numerical constant" that appears in Matveev's Theorem 2.2.14 which is $1.4 \cdot 30^{n+3} n^{4.5}$. Many celebrated Diophantine equations lead to estimates of linear forms in two or three logarithms and in these cases Matveev's Theorem 2.2.14 gives numerical constants around 10^{12} and 10^{14}, respectively.

2.4.1 A Lower Bound for $|2^m - 3^n|$

One of the simplest applications of linear forms in logarithms is to prove that $|2^m - 3^n|$ tends to infinity with $m + n$; in addition one can even get an explicit lower bound for this quantity. The following material is presented in detail in [14].

Let $n \geq 2$ be an integer and m and m' are defined by the conditions

$$2^{m'} < 3^n < 2^{m'+1}, \quad |3^n - 2^m| = \min\{3^n - 2^{m'}, 2^{m'+1} - 3^n\}.$$

Then

$$|2^m - 3^n| < 2^m, \quad (m-1)\log 2 < n\log 3 < (m+1)\log 2,$$

and the problem of finding a lower bound for $|2^m - 3^n|$ clearly reduces to this special case.

Consider the linear form

$$\Lambda = 3^n 2^{-m} - 1.$$

Applying Matveev's Theorem 2.2.14, we get

$$\log |\Lambda| > -c_0(1 + \log m).$$

It is easy to verify that we can take $c_0 = 5.87 \cdot 10^8$. Hence, the following theorem is proved.

Theorem 2.4.1 *Let m, n be positive integers. Then*

$$|2^m - 3^n| > 2^m (em)^{-5.87 \cdot 10^8}$$

This theorem enables us to find the list of all powers of 3 that increased by 5 give a power of 2.

Corollary 2.4.2 *The only integer solutions to the Diophantine equation*

$$2^m - 3^n = 5$$

are $(m, n) = (3, 1), (5, 3)$.

Applying Theorem 2.4.1, we get

$$5 > 2^m (em)^{-5.87 \cdot 10^8},$$

which implies

$$\log 5 > m \log 2 - 5.87 \cdot 10^8 (1 + \log m),$$

so that $m < 2.1 \cdot 10^{10}$ and $n < m \frac{\log 2}{\log 3} < 1.4 \cdot 10^{10}$. Moreover, the equality $2^m - 3^n = 5$ implies

$$\left| m - n \frac{\log 3}{\log 2} \right| < \frac{5}{\log 2} 3^{-n}.$$

Since

$$\frac{5}{\log 2} \cdot 3^{-n} < \frac{1}{2n}$$

for $n \geq 4$, we observe that, if (m, n) is a solution to our problem with $n \geq 4$, then $\frac{m}{n}$ is a convergent of the continued fraction expansion of $\xi = \frac{\log 3}{\log 2}$. Also, for $n < N =$

$1.4 \cdot 10^{10}$ the smallest value of $|m - n\xi|$ is obtained for the largest convergent of the continued fraction expansion of ξ with the denominator less then N. We thus get

$$\frac{5}{\log 2} \cdot 3^{-n} > \left| m - n\frac{\log 3}{\log 2} \right| > 10^{-11}, \quad 0 < n < 1.4 \cdot 10^{10}.$$

Hence, $n \leq 24$. Now, it is very easy to prove the initial statement.
    More generally, the following result can be obtained.

Theorem 2.4.3 (Bennett, [7]) *For given nonzero integers a, b, c the equation*

$$a^m - b^n = c$$

has at most two integer solutions.

2.4.2 Rep-Digit of Fibonacci Numbers

The Fibonacci sequence $(F_n)_{n \geq 0}$ is given by

$$F_0 = 0, \quad F_1 = 1, \quad \ldots, \quad F_{n+2} = F_{n+1} + F_n, \quad n \geq 0.$$

Its characteristic equation is

$$f(X) = X^2 - X - 1 = (X - \alpha)(X - \beta),$$

where $\alpha = \frac{1+\sqrt{5}}{2}$ and $\beta = \frac{1-\sqrt{5}}{2}$. We can also write

$$F_n = \frac{\alpha^n - \beta^n}{\alpha - \beta}, \quad n \geq 0.$$

    In this subsection we are concerned with those Fibonacci numbers F_n that have equal digits in base 10. Putting d for the repeated digit and assuming that F_n has m digits, the problem reduces to finding all solutions of the Diophantine equation

$$F_n = \overline{dd\ldots d}_{(10)} = d10^{m-1} + d10^{m-1} + \cdots + d = d\frac{10^m - 1}{10 - 1}, \quad d \in \{1, 2, \ldots 9\}.$$
$$(2.3)$$

Theorem 2.4.4 *The largest solution of Eq. (2.3) is $F_{10} = 55$.*

Proof Suppose that $n > 1000$. We start by proving something weaker. Our goal is to obtain some bound on n. We rewrite Eq. (2.3) as

$$\frac{\alpha^n - \beta^n}{\sqrt{5}} = d\frac{10^m - 1}{9}.$$

Next we separate *large* and *small terms* on both sides of the equation. It is easy to obtain $\alpha = \frac{-1}{\beta}$ or $\beta = \frac{-1}{\alpha}$ which implies

$$\left| \alpha^n - \frac{d\sqrt{5}}{9} 10^m \right| = \left| \beta^n - \frac{d\sqrt{5}}{9} \right| \leq |\beta^n| + \left| \frac{d\sqrt{5}}{9} \right| \leq \alpha^{-1000} + \sqrt{5} < 2.5. \quad (2.4)$$

Our goal is to get some estimates for m in terms of n. By induction on n, it is easy to prove that

$$\alpha^{n-2} < F_n < \alpha^{n-1}, \ n \geq 3.$$

Thus,

$$\alpha^{n-2} < F_n < 10^m \ \text{ or } \ n < m \frac{\log 10}{\log \alpha} + 2,$$

and

$$10^{m-1} < F_n < \alpha^{n-1}.$$

On the other hand,

$$n > \frac{\log 10}{\log \alpha}(m-1) + 1 = \frac{\log 10}{\log \alpha} m - \left(\frac{\log 10}{\log \alpha} - 1 \right) > \frac{\log 10}{\log \alpha} m - 4.$$

We deduce that

$$n \in [c_1 m - 4, \ c_1 m + 2], \qquad c_1 = \frac{\log 10}{\log \alpha} = 4.78497..$$

Since $c_1 > 4$, for $n > 1000$, we have $n \geq m$. Hence,

$$|\Lambda| = \left| \frac{d\sqrt{5}}{9} \alpha^{-n} 10^m - 1 \right| < \frac{2.5}{\alpha^n} < \frac{1}{\alpha^{n-2}},$$

which leads to

$$\log|\Lambda| = \log \frac{d\sqrt{5}}{9} - n \log \alpha + m \log 10 < -(n-2) \log \alpha.$$

Let

$$\alpha_1 = \frac{d\sqrt{5}}{9}, \ \alpha_2 = \alpha, \ \alpha_3 = 10, \ b_1 = 1, \ b_2 = -n, \ b_3 = m,$$

as well as $\mathbb{L} = \mathbb{Q}(\alpha_1, \alpha_2, \alpha_3) = \mathbb{Q}(\sqrt{5})$, so $D = 2$ and $B = n$. The minimal polynomial of α_1 over $\mathbb{Z}$ is

$$P_{\alpha_1}(X) = 81X^2 - 5d^2.$$

Hence,

$$h(\alpha_1) < \frac{1}{2}\left(\log 81 + 2\log\sqrt{5}\right) = \frac{1}{2}\log 405 < 3.01,$$

and

$$h(\alpha_2) = \frac{1}{2}(\log\alpha + 1) < 0.75, \quad h(\alpha_3) = \log 10 < 2.31.$$

We may take

$$A_1 = 6.02, \quad A_2 = 1.5, \quad A_3 = 4.62.$$

Then Matveev's Theorem 2.2.14 gives us a lower bound for Λ, namely

$$\log\Lambda > -1.4 \cdot 30^6 \cdot 3^{4.5} \cdot 4(1 + \log 4)6.02 \cdot 1.5 \cdot 4.62(1 + \log n).$$

Comparing the above inequality with

$$\log\Lambda < -(n - 2)\log\alpha,$$

we get

$$(n - 2)\log\alpha < 1.41 \cdot 30^6 \cdot 3^{4.5} \cdot 4(1 + \log 4)6.02 \cdot 1.5 \cdot 4.62(1 + \log n),$$

and

$$n < 4.5 \cdot 10^{15}.$$

Reducing the bound. Observe that the right-hand side of the inequality

$$1 - \frac{d\sqrt{5}}{9}\alpha^{-n}10^m \leq \frac{1}{\alpha^n}\left(\beta^n - \frac{d\sqrt{5}}{9}\right)$$

is negative. Writing

$$z = \log\alpha_1 - n\log\alpha_2 + m\log\alpha_3,$$

we get that

$$-\frac{2.5}{\alpha^n} < 1 - e^z < 0.$$

In particular, $z > 0$. Furthermore, we have $e^z < 1.5$ for $n > 1000$. Thus,

$$0 < e^z - 1 < \frac{2.5e^z}{\alpha^n} < \frac{4}{\alpha^n}.$$

Since $e^z - 1 > z$, we get

$$0 < m \left(\frac{\log \alpha_3}{\log \alpha_2} \right) - n + \left(\frac{\log \alpha_1}{\log \alpha_2} \right) < \frac{4}{\alpha^n \log \alpha_2} < \frac{9}{\alpha^n}.$$

We notice that

$$\left(\frac{10^m d \sqrt{5}}{\alpha^n} \right) < 2$$

and therefore

$$\alpha^n > \frac{10^m d \sqrt{5}}{2} > 10^m.$$

Hence,

$$0 < m \left(\frac{\log \alpha_3}{\log \alpha_2} \right) - n + \left(\frac{\log \alpha_1}{\log \alpha_2} \right) < \frac{9}{10^m}.$$

Since $n < 4.5 \cdot 10^{15}$, the previous inequality implies $m < 9.5 \cdot 10^{14}$. With

$$\kappa = \frac{\log \alpha_3}{\log \alpha_2}, \quad \mu = \frac{\log \alpha_1}{\log \alpha_2}, \quad A = 9, \ B = 10$$

we get

$$0 < \kappa m - n + \mu < \frac{A}{B^m},$$

where $m < N := 10^{15}$. Observe that

$$\frac{p_{35}}{q_{35}} = C_{35} = \frac{970939497358931987}{202914354378543655}$$

and $q_{35} > 202914354378543655 > 2 \cdot 10^{17} > 6N$.

For each one of the values of $d \in \{1, \ldots, 9\}$, we compute $||q_{35}\mu||$. The minimal value of this expression is obtained when $d = 5$ and is

$$0.029... > 0.02.$$

Thus, we can take $\varepsilon = 0.01 < 0.02 - 0.01 < ||q_{35}\mu|| - N||q_{35}\kappa||$. Since

$$\frac{\log(Aq_{35}/\varepsilon)}{\log B} = 21.2313...,$$

we observe that there is no solution in the range $m \in [22, 10^{15}]$. Thus, $m \leq 21$, and $n \leq 102$. However, we have assumed that $n > 1000$. To finish, we compute the values of all Fibonacci numbers modulo 10000 (their last four digits) and convince ourselves that there are no Fibonacci numbers with the desired pattern in the range $11 \leq n \leq 1000$. This example stems from [25, 36].

2.4.3   Simultaneous Pellian Equations

The following result is due to Baker and Davenport and was historically the first example of a successful use of lower bounds for linear forms in logarithms of algebraic numbers; it actually allowed the effective computation of all common members of two binary recurrent sequences with real roots; for more details see [19, 25, 36].

Theorem 2.4.5 *The only positive integer d such that $d + 1$, $3d + 1$, $8d + 1$ are all perfect squares is $d = 120$.*

Proof If $d + 1$, $3d + 1$, $8d + 1$ are all perfect squares, then we write

$$d + 1 = x^2, \ 3d + 1 = y^2, \ 8d + 1 = z^2.$$

Eliminating d from the above equations, we get

$$3x^2 - y^2 = 2, \ 8x^2 - z^2 = 7,$$

which is a *system of simultaneous Pellian equations* since it consists of two Pellian equations with a component in common. If we want x to be positive, the solutions of the above system are

$$y + x\sqrt{3} = (1 + \sqrt{3})(2 + \sqrt{3})^m,$$
$$z + x\sqrt{8} = (\pm 1 + \sqrt{8})(3 + \sqrt{8})^n,$$

where m, n are nonnegative integers. Let the sequence (v_m) be given by the recursion formula

$$v_0 = 1, \ v_1 = 3, \ v_{m+2} = 4v_{m+1} - v_m,$$

and put $x = w_n^{+,-}$, for some $n \geq 0$, where the sequences (w_n^+), (w_n^-) are defined by

$$w_0^+ = 1, \ w_1^+ = 4, \ w_{n+2}^+ = 6w_{n+1}^+ - w_n^+, \ n \in \mathbb{N},$$

$$w_0^- = 1, \ w_1^- = 2, \ w_{n+2}^- = 6w_{n+1}^- - w_n^-, \ n \in \mathbb{N}.$$

We want to solve the equation

$$v_m = w_n^{+,-}.$$

For this aim we shall use the following lemmas.

Lemma 2.4.6 *If $v_m = w_n^{+,-}$, $m, n > 2$, then*

$$0 < |\Lambda| < 7.3(2 + \sqrt{3})^{-2m},$$

where Λ is

$$\Lambda = m \log(2 + \sqrt{3}) - n \log(3 + 2\sqrt{2}) + \log \frac{2\sqrt{2}(1 + \sqrt{3})}{\sqrt{3}(2\sqrt{2} \pm 1)}.$$

Proof The expression $v_m = w_n^{+,-}$ implies

$$\frac{(1 + \sqrt{3})(2 + \sqrt{3})^m - (1 - \sqrt{3})(2 - \sqrt{3})^m}{2\sqrt{3}}$$
$$= \frac{(2\sqrt{2} \pm 1)(3 + 2\sqrt{2})^n + (2\sqrt{2} \mp 1)(3 - 2\sqrt{2})^n}{4\sqrt{2}}. \tag{2.5}$$

Obviously,

$$v_m > \frac{(1 + \sqrt{3})(2 + \sqrt{3})^m}{2\sqrt{3}},$$

$$w_n^{+,-} < \frac{(2\sqrt{2} + 1)(3 + 2\sqrt{2})^n}{2\sqrt{2}},$$

hence

$$\frac{(1 + \sqrt{3})(2 + \sqrt{3})^m}{2\sqrt{3}} < \frac{(2\sqrt{2} + 1)(3 + 2\sqrt{2})^n}{2\sqrt{2}},$$

$$(3 - 2\sqrt{2})^n < \frac{\sqrt{3}(2\sqrt{2} + 1)}{\sqrt{2}(\sqrt{3} + 1)}(2 - \sqrt{3})^m < 1.7163(2 - \sqrt{3})^m.$$

Dividing (2.5) by $\frac{2\sqrt{2} \pm 1}{4\sqrt{2}}(3 + 2\sqrt{2})^n$, we obtain

$$\left| \frac{2\sqrt{2}(1 + \sqrt{3})}{\sqrt{3}(2\sqrt{2} \pm 1)} \cdot \frac{(2 + \sqrt{3})^m}{(3 + 2\sqrt{2})^n} - 1 \right|$$
$$\leq \frac{2\sqrt{2} + 1}{2\sqrt{2} - 1}(3 - 2\sqrt{2})^{2n} + \frac{2\sqrt{2}(\sqrt{3} - 1)}{\sqrt{3}(2\sqrt{2} - 1)}(2 - \sqrt{3})^m(3 - 2\sqrt{2})^n$$
$$< \frac{2\sqrt{2} + 1}{2\sqrt{2} - 1} \cdot 1.7163^2(2 - \sqrt{3})^{2m} + \frac{2\sqrt{2}(\sqrt{3} - 1)}{\sqrt{3}(2\sqrt{2} - 1)} \cdot 1.7163(2 - \sqrt{3})^{2m}$$
$$< 7.29(2 - \sqrt{3})^{2m},$$

which proves Lemma 2.4.6.

Lemma 2.4.7 *Let $a \in \mathbb{R} \setminus \{0\}, a > 1$. If $|x| < a$, then*

$$|\log(1 + x)| < \frac{-\log(1 - a)}{a}|x|. \tag{2.6}$$

Proof We observe that the function

$$\frac{\log(1 + x)}{x}$$

is positive and strictly decreasing for $|x| < 1$. Consequently, for $|x| < a$, Inequality (2.6) holds for $x = -a$.

We shall investigate the following linear form in three logarithms:

$$\Lambda = m \log(2 + \sqrt{3}) - n \log(2\sqrt{2} + 3) + \log \frac{2\sqrt{2}(\sqrt{3} + 1)}{\sqrt{3}(2\sqrt{2} \pm 1)}.$$

Let

$$\alpha_1 = 2 + \sqrt{3}, \quad \alpha_2 = 3 + 2\sqrt{2}, \quad \alpha_3 = \frac{2\sqrt{2}(1 + \sqrt{3})}{\sqrt{3}(2\sqrt{2} \pm 1)},$$

$$b_1 = m, \quad b_2 = -n, \quad b_3 = 1, \quad D = [\mathbb{Q}(\alpha_1, \alpha_2, \alpha_3) : \mathbb{Q}] = 4.$$

The minimal polynomials over $\mathbb{Z}$ are

$$P_{\alpha_1}(x) = x^2 - 4x + 1,$$
$$P_{\alpha_2}(x) = x^2 - 6x + 1,$$
$$P_{\alpha_3}(x) = 441x^4 - 2016x^3 + 2880x^2 - 1536x + 256,$$

hence

$$h''(\alpha_1) = \frac{1}{2} \log(2 + \sqrt{3}) < 0.6585,$$

$$h''(\alpha_2) = \frac{1}{2} \log(3 + 2\sqrt{2}) < 0.8814,$$

and

$$h''(\alpha_3) = \frac{1}{4} \log\left(441 \frac{2(4 + \sqrt{2})(3 + \sqrt{3})}{21} \frac{2(4 - \sqrt{2})(3 + \sqrt{3})}{21}\right) < 1.7836.$$

Applying the Baker–Wüstholz Theorem 2.2.13, we get a lower bound for Λ, namely

$$\log|\Lambda| \geq -3.96 \cdot 10^{15} \log m.$$

According to Lemma 2.4.6, we may conclude that

$$m < 6 \cdot 10^{16}.$$

This upper bound is rather big so we reduce it using Baker–Davenport reduction. Applying Lemma 2.4.6, we get the upper bound

$$\Lambda = m \log(2 + \sqrt{3}) - n \log(2\sqrt{2} + 3) + \log \frac{2\sqrt{2}(\sqrt{3} + 1)}{\sqrt{3}(2\sqrt{2} \pm 1)} < 7.29(2 + \sqrt{3})^{-2m}.$$

Let

$$N = 6 \cdot 10^{16}, \quad \kappa = \frac{\log \alpha_1}{\log \alpha_2}, \quad \mu = \frac{\log \alpha_3}{\log \alpha_2}, \quad A = \frac{7.3}{\log \alpha_2}, \quad B = (2 + \sqrt{3})^2.$$

The next step is to find a convergent $\frac{p}{q}$ of the continued fraction expansion of $\kappa = \frac{\log \alpha_1}{\log \alpha_2}$ such that $q > 6N$. The first such convergent is

$$\frac{p}{q} = \frac{7422659006396841111}{9935223607325971120}.$$

Before calculating ε, we observe that

$$||\kappa q||N \approx 0.0187822, \qquad ||\mu q|| \approx 0.00762577.$$

Unfortunately,

$$\varepsilon = ||\mu q|| - ||\kappa q||N < 0.$$

If we want to use Baker–Davenport reduction, we have to find another convergent $\frac{p}{q}$ of the continued fraction expansion for which the condition $\varepsilon > 0$ is satisfied.

The next convergent $\frac{p}{q}$ of κ that satisfies condition $q > 6N$ is

$$\frac{p}{q} = \frac{22975706401873543392}{30752966078889333649}.$$

It follows that

$$\varepsilon = ||\mu q|| - ||\kappa q||N \approx 0.296651 > 0.$$

The given inequality does not have any solutions in integers m such that

$$\frac{\log \left(\frac{Aq}{\varepsilon} \right)}{\log B} \le m < N.$$

Thus, the new upper bound for m is 17. Repeating the procedure once again, we get $m \le 4$, and there are only two solutions, namely

$$v_0 = w_0^{+,-} = 1,$$

which is the trivial solutions of our Diophantine equation with $d = 0$, and

$$v_2 = w_2^- = 11,$$

which suits the case when $d = 120$.

2.4.4 Fibonacci Numbers and the Property of Diophantus

The next result is due to Dujella [17]. He proved that, if k and d are positive integers such that the set

$$\{F_{2k}, F_{2k+2}, F_{2k+4}, d\}$$

is a $D(1)$-quadruple, then $d = 4F_{2k+1}F_{2k+2}F_{2k+3}$, where F_k is k-th Fibonacci number. This is a generalization of the Theorem 2.4.5 of Baker and Davenport for $k = 1$. Here a set $\{a_1, a_2, a_3, a_4\}$ of distinct positive integers is called *a $D(1)$-quadruple*, if $a_i a_j + n$ is a perfect square for every i, j with $1 \leq i < j \leq 4$.

Proof (Sketch) Let $k \geq 2$ be a positive integer and

$$a = F_{2k}, \quad b = F_{2k+2}, \quad c = F_{2k+4}.$$

Then $c = 3b - a$. Furthermore,

$$ab + 1 = (b - a)^2, \quad ac + 1 = b^2, \quad bc + 1 = (a + b)^2.$$

If we assume that d is a positive number such that $\{a, b, c, d\}$ has the property $D(1)$ of Diophantus, it implies that there exist positive integers x, y, z such that

$$ad + 1 = x^2, \quad bd + 1 = y^2, \quad (3b - a)d + 1 = z^2.$$

Eliminating d, we get a system of Pellian equations

$$ay^2 - bx^2 = a - b, \quad az^2 - (3b - a)x^2 = 2a - 3b. \qquad (2.7)$$

Dujella proved the following lemmas.

Lemma 2.4.8 *Let x, y, z be positive integer solutions of the system of Pellian equations (2.7). Then there exist integers m and n such that*

$$x = v_m = w_n,$$

where (v_m) is given by

$$v_0 = 1, \quad v_1 = b, \quad v_{m+2} = 2(b - a)v_{m+1} - v_m, \quad m \in \mathbb{Z},$$

and the two-sided sequence (w_n) is defined by

$$w_0 = 1, \quad w_1 = a + b, \quad w_{n+2} = 2bw_{n+1} - w_n, \quad n \in \mathbb{Z}.$$

In order to apply Baker's method, it is convenient to consider the two-sided sequence as two ordinary sequences. Therefore, instead of the sequence $(v_m)_{m \in \mathbb{Z}}$, Dujella considered two sequences $(v_m)_{m \geq 0}$ and $(v_m)_{m \leq 0}$ and applied the same method for the sequence $(w_n)_{n \in \mathbb{Z}}$. Thus, four equations of the form

$$v_m = w_n$$

have to be considered.

Lemma 2.4.9 If $v_m = w_n$, and $m \neq 0$, then

$$0 < m \log(b - a + \sqrt{ab}) - n \log(b + \sqrt{ac}) +$$
$$+ \log \frac{\sqrt{c}(\pm\sqrt{a} + \sqrt{b})}{\sqrt{b}(\pm\sqrt{a} + \sqrt{c})} < 4(b - a + \sqrt{ab})^{-2m}.$$

In the present situation, $l = 3$, $d = 4$, $B = m$ and

$$\alpha_1 = b - a + \sqrt{ab}, \ \alpha_2 = b + \sqrt{ac}, \ \alpha_3 = \frac{\sqrt{c}(\pm\sqrt{a} + \sqrt{b})}{\sqrt{b}(\pm\sqrt{a} + \sqrt{c})},$$

$$h'(\alpha_1) = \frac{1}{2} \log \alpha_1 < 1.05 \log, \ h'(\alpha_2) = \frac{1}{2} \log \alpha_2 < 1.27 \log a,$$

$$h'(\alpha_3) = \frac{1}{4} \log(bc(c - a)(\sqrt{a} + \sqrt{b})^2) < 2.52 \log a,$$

$$\log 4(b - a + \sqrt{ab})^{-2m} < \log a^{-2m} = -2m \log a.$$

Hence,

$$2m \log a < 3.822 \cdot 10^{15} \cdot 3.361 \log^3 a \log m,$$

and

$$\frac{m}{\log m} < 6.423 \cdot 10^{15} \log^2 a. \tag{2.8}$$

Applying Lemma 2.4.8, we get

$$|m| \geq 2b - 2 > 4a. \tag{2.9}$$

Comparing (2.8) and (2.9), we obtain

$$\frac{m}{\log^3 m} < 6.423 \cdot 10^{15},$$

which implies $m < 8 \cdot 10^{20}$, $a = F_{2k} < 2 \cdot 10^{20}$. The author has proved the theorem for $k \geq 49$.

It remains to prove the theorem for $2 \leq k \leq 48$. Dujella used Lemma 2.3.1 with

$$\kappa = \frac{\log \alpha_1}{\log \alpha_2}, \quad \mu = \frac{\log \alpha_3}{\log \alpha_2}, \quad A = \frac{4}{\log \alpha_3}, \quad B = (b - a + \sqrt{ab})^2$$

as well as $N = 8 \cdot 10^{20}$ and gets a new bound $m \leq N_0$, where $N_0 \leq 12$. Repeating the method one more time, he obtained a new upper bound $m \leq 2$ which completes the proof.

2.4.5 The Non-extendibility of Some Parametric Families of $D(-1)$-Triples

Definition 2.1 Let n be a nonzero integer. A set $\{a_1, \ldots, a_m\}$ of m distinct positive integers is called *a Diophantine m-tuple with the property* $D(n)$, or simply *a* $D(n)$-*m-tuple*, if $a_i a_j + n$ is a perfect square for any i, j with $1 \leq i < j \leq m$.

The set $\{1, 3, 8, 120\}$ considered in the previous Sect. 2.4.3 is known as the first example of a $D(1)$-quadruple found by Fermat.[28] In 1969, Baker and Davenport proved that $\{1, 3, 8\}$ cannot be extended to a $D(1)$-quintuple (see [5]). This result was generalized by Dujella [15], who showed that the $D(1)$-triple $\{k - 1, k + 1, 4k\}$ for an integer k cannot be extended to a $D(1)$-quintuple, and by Dujella and Pethő [22], who proved that the $D(1)$-pair $\{1, 3\}$ cannot be extended to a $D(1)$-quintuple. It is a conjectured that there does not exist a $D(1)$-quintuple. The most general results on this conjecture are due to Dujella [18] who proved that there does not exist a $D(1)$-sextuple and that there exist at most finitely many $D(1)$-quintuples. There have been some improvements on those results recently, but the conjecture still remains open.

In contrast to the case $n = 1$, it is conjectured that there does not exist a $D(-1)$-quadruple (see [16]). The first important step in this direction was done by Dujella and Fuchs [21] who showed that, if $\{a, b, c, d\}$ is a $D(-1)$-quadruple with $a < b < c < d$, then $a = 1$. Later Dujella, Filipin and Fuchs [23] showed that there exist at most finitely many $D(-1)$-quadruples. The number of $D(-1)$-quadruples is now known to be bounded by $5 \cdot 10^{60}$ (see [24]). However, this bound is too large for verifying the conjecture by present day computers. Recently, He and Togbé [29] proved that the $D(-1)$-triple $\{1, k^2 + 1, k^2 + 2k + 2\}$ cannot be extended to a $D(-1)$-quadruple. Their result and the proof appears to be very important because of their use of a linear form in two logarithms (instead of three) for the first time; this leads to a much better upper bound for the solutions which shortens the reduction time significantly. In this subsection, we extend their method and apply it to several other families of $D(-1)$-triples. Let us also mention that it is not always possible to use linear forms in two logarithms. In the sequel we only explain the idea and give a sketch of the proofs; more details can be found in [26].

---

[28]Pierre de Fermat (1601–1665), a French mathematician.

Introduction

Let $\{1, b, c\}$ be a $D(-1)$-triple with $b < c$. We define positive integers r, s and t by

$$b - 1 = r^2, \ c - 1 = s^2, \ bc - 1 = t^2.$$

Then, s and t satisfy

$$t^2 - bs^2 = r^2. \tag{2.10}$$

It can be proven that Diophantine equation (2.10) has at least three classes of solutions belonging to

$$(t_0, s_0) = (r, 0), \ (b - r, \pm(r - 1))$$

(see [27, p. 111]). We call a positive solution (t, s) of (2.10) *regular* if (t, s) belongs to one of these three classes. However, it is possible for a solution (t, s) not to be regular. In general, we do not know in advance how many classes of solutions we have, except for some special type of b.

Remark 2.4.10 An example having non-regular solutions can be found in the case of $r = 2q^2$, where q is a positive integer. Then (2.10) has two more classes of solutions belonging to

$$(t_0', s_0') = (2q^3 + q, \pm q).$$

Our goal is to prove the following theorem.

Theorem 2.2 *Let (t, s) be a regular solution of (2.10) and let $c = s^2 + 1$. Then, the system of Diophantine equations*

$$y^2 - bx^2 = r^2,$$
$$z^2 - cx^2 = s^2$$

has only trivial solutions $(x, y, z) = (0, \pm r, \pm s)$. Furthermore, if $r = 2q^2$ for some positive integer q, then the same is true for any positive solution (t, s) of (2.10) belonging to the same class as one of $(2q^3 + q, \pm q)$.

By [23, Theorem 1] we have that $c < 11b^6$ (using the hyper-geometric method), hence the above-mentioned result of He and Togbé shows that it is enough to prove Theorem 2.2 for $c = c_i$ with $2 \leq i \leq 7$, where

$$c_2 = 4r^4 + 1,$$
$$c_3 = (4r^3 - 4r^2 + 3r - 1)^2 + 1,$$
$$c_4 = (4r^3 + 4r^2 + 3r + 1)^2 + 1,$$
$$c_5 = (8r^4 + 4r^2)^2 + 1,$$
$$c_6 = (16r^5 - 16r^4 + 20r^3 - 12r^2 + 5r - 1)^2 + 1,$$
$$c_7 = (16r^5 + 16r^4 + 20r^3 + 12r^2 + 5r + 1)^2 + 1$$

(see [27, p. 111]), and in the case of $r = 2q^2$, additionally for $c = c_i'$ with $1 \leq i \leq 5$, where

$$c_1' = (4q^3 - q)^2 + 1,$$
$$c_2' = (16q^5 + 4q^3 + q)^2 + 1,$$
$$c_3' = (64q^7 - 16q^5 + 8q^3 - q)^2 + 1,$$
$$c_4' = (256q^9 + 64q^7 + 48q^5 + 8q^3 + q)^2 + 1,$$
$$c_5' = (1024q^{11} - 256q^9 + 256q^7 - 48q^5 + 12q^3 - q)^2 + 1.$$

It is easy to see that Theorem 2.2 immediately implies

Corollary 2.4.11 *Let (t, s) be either a regular solution of (2.10) or, in the case of $r = 2q^2$ for some positive integer q, a regular solution or a positive solution of (2.10) belonging to the same class as one of $(2q^3 + q, \pm q)$. Let $c = s^2 + 1$. Then, the $D(-1)$-triple $\{1, b, c\}$ cannot be extended to a $D(-1)$-quadruple.*

It was proven in [27, p. 111], that if r is prime, then (2.10) has only regular solutions. We can generalize this to find that, if $r = p^k$ or $2p^k$ for an odd prime p and a positive integer k, then (2.10) has only regular solutions, except in the case of $r = 2p^k$ with k even. In the latter case, there are exactly five classes of solutions. Furthermore, if $b = p$ or $2p^k$, then (2.10) has only regular solutions ($b = p^k$ can occur only if $k = 1$, since $b = r^2 + 1$; [35]). Hence, we get another corollary of the Theorem 2.2.

Corollary 2.4.12 *Let r be a positive integer and let $b = r^2 + 1$. Suppose that one of the following assumptions holds for an odd prime p and a positive integer k:*
(i) $b = p$; (ii) $b = 2p^k$; (iii) $r = p^k$; (iv) $r = 2p^k$.
Then, the system of Diophantine equations

$$y^2 - bx^2 = r^2,$$
$$z^2 - cx^2 = s^2$$

has only the trivial solutions $(x, y, z) = (0, \pm r, \pm s)$, where (t, s) is a positive solution of (2.10) and $c = s^2 + 1$. Moreover, the $D(-1)$-pair $\{1, b\}$ cannot be extended to a $D(-1)$-quadruple.

The System of Pellian Equations

Let $\{1, b, c\}$ be a $D(-1)$-triple with $b < c$, and let r, s, t be positive integers defined by $b - 1 = r^2, c - 1 = s^2, bc - 1 = t^2$. Suppose that we can extend the triple $\{1, b, c\}$ to a $D(-1)$-quadruple with element d. Then, there exist integers x, y, z such that

$$d - 1 = x^2, \quad bd - 1 = y^2, \quad cd - 1 = z^2.$$

Eliminating d, we obtain the system of simultaneous Diophantine equations

$$z^2 - cx^2 = s^2, \tag{2.11}$$

$$bz^2 - cy^2 = c - b, \tag{2.12}$$

$$y^2 - bx^2 = r^2. \tag{2.13}$$

We may assume that $c < 11b^6$ (according to Theorem 1, [23]). The positive solutions (z, x) of Eq. (2.11) and (z, y) of Eq. (2.12) are respectively given by

$$z + x\sqrt{c} = s(s + \sqrt{c})^{2m} \quad (m \geq 0),$$

$$z\sqrt{b} + y\sqrt{c} = (s\sqrt{b} \pm r\sqrt{c})(t + \sqrt{bc})^{2n} \quad (n \geq 0)$$

(see Lemmas 1 and 5 in [23]). Using that y is a common solution of Eqs. (2.12) and (2.13), He and Togbé proved that the positive solutions (y, x) of Eq. (2.13) are given by

$$y + x\sqrt{b} = r(r + \sqrt{b})^{2l}, \quad l \geq 0,$$

and, moreover, they proved the following proposition.

Proposition 2.4.13 ([29, Proposition 2.1]) *The $D(-1)$-triple $\{1, b, c\}$ can be extended to a $D(-1)$-quadruple if and only if the system of simultaneous Pellian equations*

$$(z/s)^2 - c(x/s)^2 = 1,$$

$$(y/r)^2 - b(x/r)^2 = 1$$

has a positive integer solution (x, y, z).

Proposition 2.4.13 implies that we can write $x = sv_m = ru_l$, where

$$v_m = \frac{\alpha^{2m} - \alpha^{-2m}}{2\sqrt{c}} \quad \text{and} \quad u_l = \frac{\beta^{2l} - \beta^{-2l}}{2\sqrt{b}}$$

are positive solutions of the Pellian equations $Z^2 - cX^2 = 1$ and $Y^2 - bW^2 = 1$, respectively, where $\alpha = s + \sqrt{c}$ and $\beta = r + \sqrt{b}$. We have mentioned before that we cannot always use linear forms in two logarithms. More precisely, for our approach

α and β should be near to each other or one of them has to be near to a power of the other one.

Gap Principles

We next consider the extension of $D(-1)$-triples $\{1, b, c\}$ with

$$c = c_2, c_3, c_4, c_5, c_6, c_7, c_1', c_2', c_3', c_4', c_5'$$

from above. We shall establish gap principles for these special cases.
Let us define the linear form Λ in three logarithms

$$\Lambda = 2m \log \alpha - 2l \log \beta + \log \frac{s\sqrt{b}}{r\sqrt{c}}.$$

The proofs of the following lemmas can be found in [29]; the results rely on Baker's theory of linear forms in logarithms.

Lemma 2.4.14 ([29, Lemma 3.1]) *If* $sv_m = ru_l$ *has a solution with* $m \neq 0$, *then*

$$0 < \Lambda < \frac{b}{b-1} \cdot \beta^{-4l}.$$

Lemma 2.4.15 ([29, Lemma 3.3]) *If* $sv_m = ru_l$ *has a solution with* $m \neq 0$, *then* $m \log \alpha < l \log \beta$.

The proofs of the following lemmas can be found in [26]; these results are based on the property that an algebraic number α is close to some power of β.

Lemma 2.4.16 *Let* $c = c_2 = 4r^4 + 1$. *If the equation* $sv_m = ru_l$ *has a solution with* $m \neq 0$, *then*

$$m > \frac{\Delta}{2} \cdot \alpha \log \beta,$$

where Δ *is a positive integer.*

Lemma 2.4.17 *Let* $c = c_3 = (4r^3 - 4r^2 + 3r - 1)^2 + 1$ *or* $c = c_4 = (4r^3 + 4r^2 + 3r + 1)^2 + 1$. *If the equation* $sv_m = ru_l$ *has a solution with* $m \neq 0$, *then*

$$m > \frac{3\Delta - 1}{3} \cdot \frac{8r}{9} \log \beta,$$

where Δ *is a positive integer.*

Lemma 2.4.18 *Let* $c = c_1' = (4q^3 - q)^2 + 1$. *If the equation* $sv_m = ru_l$ *has a solution with* $m \neq 0$, *then*

$$m > \frac{\Delta}{6} \cdot \alpha \log \beta,$$

where Δ is a positive integer.

Similar lemmas for other choices of c can be obtained. The following table contains information how to choose Δ and a lower bound for m.

c	Δ	a lower bound for m
c_5	$4m - l$	$m > \Delta \cdot \frac{8r^4}{4r^2+1} \log \beta$
c_6	$5m - l$	$m > \frac{5\Delta-1}{5} \cdot \frac{32r}{33} \log \beta$
c_7	$l - 5m$	$m > \frac{5\Delta-1}{5} \cdot \frac{32r}{33} \log \beta$
c'_2	$2l - 5m$	$m > \frac{10\Delta-4}{5} \cdot q^2 \log \beta$
c'_3	$7m - 2l$	$m > \frac{3\Delta}{2} \cdot q^2 \log \beta$
c'_4	$2l - 9m$	$m > \frac{18\Delta-4}{9} \cdot q^2 \log \beta$
c'_5	$11m - 2l$	$m > \frac{3\Delta}{2} \cdot q^2 \log \beta$

Linear Forms in Two Logarithms

Next we shall apply the following result due to Laurent,[29] M. Mignotte[30] and Y. Nesterenko[31] to our linear form Λ.

Lemma 2.4.19 ([33, Corollary 2]) *Let γ_1 and γ_2 be multiplicatively independent, positive algebraic numbers, $b_1, b_2 \in \mathbb{Z}$ and*

$$\Lambda = b_1 \log \gamma_1 + b_2 \log \gamma_2.$$

Let $D := [\mathbb{Q}(\gamma_1, \gamma_2) : \mathbb{Q}]$, for $i = 1, 2$ let

$$h_i \geq \max \left\{ h(\gamma_i), \frac{|\log \gamma_i|}{D}, \frac{1}{D} \right\},$$

where $h(\gamma)$ is the absolute logarithmic height of γ, and

$$b' \geq \frac{|b_1|}{Dh_2} + \frac{|b_2|}{Dh_1}.$$

If $\Lambda \neq 0$, then

$$\log |\Lambda| \geq -24.34 \cdot D^4 \left(\max \left\{ \log b' + 0.14, \frac{21}{D}, \frac{1}{2} \right\} \right)^2 h_1 h_2.$$

---

[29]Michel Laurent, a French mathematician.

[30]Maurice Mignotte, a French mathematician.

[31]Yuri Valentinovich Nesterenko (1946), a Soviet and Russian mathematician.

This lemma has also been used by He and Togbé in [29]. We are dealing with the same linear form, but only with a different c. Using the same method to transform our form to a linear form in two logarithms, then applying Lemma 2.4.19 for $c = c_2, c_3, \ldots, c_7, c_1', \ldots, c_5'$, and combining the lower bound for $|\Lambda|$ together with the gap principles, we can prove Theorem 2.2 for large values of r. That γ_1 and γ_2 are multiplicatively independent follows from the fact that α and β are multiplicatively independent algebraic units and $\frac{r\sqrt{c}}{s\sqrt{b}}$ is not an algebraic unit.

As an example we consider the case $c = c_2 = 4r^4 + 1$. We can write

$$\Lambda = 2m \log\left(\frac{\alpha}{\beta^2}\right) - \log\left(\beta^{-2\Delta} \cdot \frac{r\sqrt{c}}{s\sqrt{r^2 + 1}}\right),$$

where $\Delta = 2m - l$ is defined as in Lemma 2.4.16. In the notation of Lemma 2.4.19 we have

$$D = 4, \ b_1 = 2m, \ b_2 = -1, \ \gamma_1 = \frac{\alpha}{\beta^2}, \ \gamma_2 = \beta^{-2\Delta} \cdot \frac{r\sqrt{c}}{s\sqrt{b}}.$$

Furthermore,

$$h(\gamma_1) \le h\left(\frac{\alpha}{\beta}\right) + h(\beta) = \frac{1}{2}\log\alpha + \frac{1}{2}\log\beta < \log\alpha,$$

hence, for h_1, we can take $h_1 = \log\alpha$. Moreover,

$$h\left(\frac{r\sqrt{c}}{s\sqrt{b}}\right) = \frac{1}{2}\log((c-1)b) < \frac{1}{2}\log\beta^6 = 3\log\beta,$$

which yields

$$h(\gamma_2) < (\Delta + 3)\log\beta =: h_2.$$

For $r \ge 10$, we find

$$\frac{|b_2|}{Dh_1} = \frac{1}{4\log\alpha} < 0.042,$$

and then

$$b' = \frac{m}{2(\Delta + 3)\log\beta} + 0.042.$$

Now Lemma 2.4.16 implies

$$\frac{m}{2(\Delta + 3)\log\beta} > \frac{\Delta}{4(\Delta + 3)} \cdot \alpha \ge \frac{\alpha}{16} > 169$$

for $r \ge 26$. Thus, for $r \ge 26$, we get $\log b' + 0.14 > \frac{21}{D}$ and applying Lemma 2.4.19 we conclude

$$\log |\Lambda| \geq -24.34 \cdot 4^4 \cdot (\log b' + 0.14)^2 \cdot \log \alpha \cdot (\Delta + 3) \log \beta.$$

On the other hand, Lemma 2.4.14 yields

$$\log |\Lambda| < 0.002 - 4l \log \beta.$$

Combining these lower and upper bounds for Λ, we obtain

$$\frac{l}{\log \alpha} < \frac{0.002}{4 \log \alpha \log \beta} + 24.34 \cdot 64 (\log b' + 0.14)^2 (\Delta + 3).$$

Furthermore, $m \log \alpha < l \log \beta$ gives us

$$\frac{m}{2(\Delta + 3) \log \beta} < 0.0001 + 24.34 \cdot 32 (\log b' + 0.14)^2$$

and finally

$$b' < 0.042 + 778.88 (\log b' + 0.14)^2,$$

which implies $b' < 106996$. It furthermore gives $m < 213992(\Delta + 3) \log \beta$ and

$$\alpha < \frac{2m}{\Delta \log \beta} < 427983 \cdot \frac{\Delta + 3}{\Delta} < 1.72 \cdot 10^6,$$

from which we deduce $r < 656$. Thus, we have proved Theorem 2.2 for $c = c_2$ and $r \geq 656$.

The cases $c = c_3, c_4 = (4r^3 \mp 4r^2 + 3r \mp 1)^2 + 1$ are described in details in [26]. The upper bounds for r and q in the remaining cases are given in the following table.

c	an upper bound for r or q
c_3	$r < 1.81 \cdot 10^6$
c_4	$r < 1.81 \cdot 10^6$
c_5	$r < 802$
c_6	$r < 1.94 \cdot 10^6$
c_7	$r < 1.94 \cdot 10^6$
c_2'	$q < 846$
c_3'	$q < 846$
c_4'	$q < 949$
c_5'	$q < 1000$

The Reduction Method and the Proof of Theorem 2.2

We have just proven Theorem 2.2 for large parameters r and q. We are left to consider the cases of small r and q. Using Baker–Davenport reduction, it turns out that in all remaining cases there is no extension of the triple $\{1, b, c\}$ to a quadruple $\{1, b, c, d\}$.

Some useful results from [23] can be used. We know that, if we have the extension of our triple with the element d, then $cd - 1 = z^2$, where $z = V_m = W_n$ such that

$$V_0 = s, \quad V_1 = (2c - 1)s, \quad V_{m+2} = (4c - 2)V_{m+1} - V_m$$

and

$$W_0 = s, \quad W_1 = (2bc - 1)s \pm 2rtc, \quad W_{n+2} = (4bc - 2)W_{n+1} - W_n.$$

We use the following lemmas.

Lemma 2.4.20 ([23, Lemma 11]) *If* $V_m = W_n$, $n \neq 0$, *then*

$$0 < 2n \log(t + \sqrt{bc}) - 2m \log(s + \sqrt{c}) + \log \frac{s\sqrt{b} \pm r\sqrt{c}}{2\sqrt{b}} < (3.96bc)^{-n+1}.$$

From the proofs of Propositions 2, 3, 4 in [23] we know that $n < 10^{20}$ in all cases. Applying Baker–Davenport reduction with

$$\kappa = \frac{\log(t + \sqrt{bc})}{\log(s + \sqrt{c})}, \quad \mu = \frac{\log \frac{s\sqrt{b} \pm r\sqrt{c}}{2\sqrt{b}}}{2\log(s + \sqrt{c})}, \quad A = \frac{3.96bc}{2\log(s + \sqrt{c})}, \quad B = 3.96bc$$

and $N = 10^{21}$ with any choice of r and c left, we get after two steps that $n < 2$. Here, one may also use that $D(-1)$-triples $\{1, b, c\}$ cannot be extended to a quadruple for $r \leq 143000$, which was verified by computer. Hence, in some cases one can avoid to use reduction at all.

We are still left to deal with the cases of small indices m and n in the equation $z = V_m = W_n$. From [21] we know that $n \geq 3$; otherwise we have only the trivial solution (corresponding with an extension with $d = 1$, which is no real extension, because we ask for elements in $D(n)$-m-tuple to be distinct).

The following lemma, which was proved in [26], and which examines the fundamental solutions of (2.10) in the cases of $b = p$, $2p^k$ and $r = p^k$, $2p^k$, together with Theorem 2.2 implies Corollary 2.4.12.

Lemma 2.4.21

(1) *If* $b = p$ *or* $2p^k$ *for an odd prime* p *and a positive integer* k, *then Diophantine Equation (2.10) has only regular solutions.*

(2) *If* $r = p^k$ *or* $2p^k$ *for an odd prime* p *and a positive integer* k, *then Diophantine Equation (2.10) has only regular solutions, except in the case of* $r = 2p^{2i}$ *with* i *a positive integer, where it in addition has exactly two classes of solutions belonging to* $(2p^{3i} + p^i, \pm p^i)$.

2.4.6 Pure Powers in Binary Recurrent Sequences

The Lucas numbers $(L_n)_{n \geq 0}$ are given by

$$L_0 = 2, \ L_1 = 1, \ \ldots, \ L_{n+2} = L_{n+1} + L_n, \ n \geq 0.$$

Recall that Fibonacci numbers (F_n) as well as Lucas numbers (L_n) are defined by

$$F_n = \frac{\alpha^n - \beta^n}{\sqrt{5}}, \ L_n = \alpha^n + \beta^n, \ \alpha = \frac{1 + \sqrt{5}}{2}, \ \beta = \frac{1 - \sqrt{5}}{2},$$

respectively. Now suppose

$$F_n = y^p$$

is a pure power. Since

$$\alpha^n - \sqrt{5} y^p = O(\alpha^{-n}),$$

we find

$$\Lambda = n \log \alpha - p \log y - \log \sqrt{5} = O(\alpha^{-2n}) = O(y^{-2p}).$$

There exist integers k, r such that $n = kp + r$ with $|r| \leq \frac{p}{2}$, hence we have

$$\Lambda = p \log \left(\frac{\alpha^k}{y} \right) + r \log \alpha - \log \sqrt{5}$$

which is a linear form in three logarithms. If we apply Matveev's Theorem 2.2.14, we get

$$\log |\Lambda| \geq -c^* \log y \log p.$$

Comparing both estimates of $|\Lambda|$, we see that the exponent p is bounded. Matveev's Theorem 2.2.14 implies $p < 3 \cdot 10^{13}$, but a special estimate for linear forms in three logarithms implies the sharper upper bound is $p < 2 \cdot 10^8$ which is suitable for computer calculations.

For Lucas numbers a similar study leads to a linear form in two logarithms and $p < 300$ provided that $L_n = y^p$. By this reasoning, it can be proved [10] that all perfect powers in the Fibonacci and Lucas sequences are

$$F_0 = 0, \ F_1 = F_2 = 1, \ F_6 = 8 = 2^3, \ F_{12} = 144 = 12^2;$$

$$L_1 = 1, \ L_3 = 4 = 2^2.$$

Using this method we can solve many similar problems, for example, all Pell numbers for which $P_n + 4$ is a perfect square are given by $P_0 = 0$, $P_3 = 5$ and $P_4 = 12$. Recall that the Pell numbers are given by the recursion

$$P_0 = 0, \ P_1 = 1, \ \ldots, \ P_{n+2} = 2P_{n+1} + P_n, \ n \geq 0.$$

2.4.7 Lucas Numbers and the Biggest Prime Factor

Next we are interested in finding all Lucas numbers for which the biggest prime factor is less than or equal to 5.

We may express Lucas numbers as

$$L_n = \left(\frac{1+\sqrt{5}}{2}\right)^n + \left(\frac{1-\sqrt{5}}{2}\right)^n, \quad n \in \mathbb{N}.$$

For Lucas numbers the length of the period of the sequence $(L_n \bmod 5)$ is equal to 4 with the cycle $\{1, 3, 4, 2\}$; therefore 5 can never be a divisor of any Lucas number. We want to find all Lucas numbers such that

$$2^k 3^l = \left(\frac{1+\sqrt{5}}{2}\right)^n + \left(\frac{1-\sqrt{5}}{2}\right)^n, \quad n, k, l, m \in \mathbb{N}.$$

Let $\alpha = \frac{1+\sqrt{5}}{2}$, so the previous expression can be rewritten as

$$2^k 3^l = \alpha^n - \alpha^{-n}, \quad k, l, n \in \mathbb{N}.$$

Thus the corresponding linear form in logarithms is

$$\Lambda = 2n \log \alpha - k \log 2 - l \log 3.$$

Now, applying Matveev's Theorem 2.2.14, we have $n = 3$, $D = 2$, and

$$A_1 \geq h'(2) = \max\{2 \log 2, |\log 2|, 0.16\} = 1.38 < 2,$$

$$A_2 \geq h'(3) = \max\{2 \log 3, |\log 3|, 0.16\} = 2.19 < 3,$$

$$A_3 \geq h'\left(\frac{1+\sqrt{5}}{2}\right) = \max\{2 \log \frac{1+\sqrt{5}}{2}, \left|\log \frac{1+\sqrt{5}}{2}\right|, 0.16\} = 0.48 < 1.$$

We get

$$\log |\Lambda| \geq -7.28022 \cdot 10^{15}(1 + \log 2n).$$

After finding the upper bound from the expression

$$\frac{L_n}{\alpha^n} - 1 = -\alpha^{-2n},$$

and

$$\log |\Lambda| < -2n \log \frac{1+\sqrt{5}}{2},$$

we find that $n < 3.17654 \cdot 10^{17}$. After applying Baker–Davenport reduction, we get $n < 14$, so we may conclude that the Lucas numbers for which the biggest prime factors are less or equal to 5 are

$$L_0 = 2, \quad L_2 = 3, \quad L_4 = 4 = 2^2, \quad L_6 = 18 = 2 \cdot 3^2.$$

2.4.8 Pillai's Equation

Given positive integers $a > b > 1$, Pillai[32] [42] proved that there are only finitely many integers $c \neq 0$ admitting more than one representation of the form

$$c = a^x - b^y$$

in nonnegative integers x, y. In particular, the equation

$$a^x - b^y = a^{x_1} - b^{y_1}, \quad \text{with } (x, y) \neq (x_1, y_1) \tag{2.14}$$

has only finitely many integer solutions. We shall apply the technique of lower bounds for linear forms in logarithms of algebraic numbers to find all the solutions for

$$(a, b) = (3, 2).$$

Proposition 2.4.22 *The only nontrivial solutions of Eq. (2.14) with $(a, b) = (3, 2)$ are*

$$3^1 - 2^2 = 3^0 - 2^1, \quad 3^2 - 2^4 = 3^0 - 2^3, \quad 3^2 - 2^3 = 3^1 - 2^1,$$

$$3^3 - 2^5 = 3^1 - 2^3, \quad 3^5 - 2^8 = 3^1 - 2^4.$$

Proof The initial equation can be rewritten as

$$3^x - 3^{x_1} = 2^y - 2^{y_1}.$$

After relabeling the variables, we may assume that $x > x_1$. Consequently, $y > y_1$. Since

$$2 \cdot 3^{x-1} = 3^x - 3^{x-1} \leq 3^x - 3^{x_1} = 2^y - 2^{y_1} < 2^y,$$

we get $x < y$. Let $B = y$. Now,

$$3^{x_1} \mid (2^y - 2^{y_1}) = 2^{y_1}(2^{y-y_1} - 1).$$

We observe that $3^m \mid (2^n - 1)$ if and only if $2 \cdot 3^{m-1} \mid n$. In particular,

---

[32]Subbayya Sivasankaranarayana Pillai (1901–1950), an Indian mathematician.

$$x_1 \leq 1 + \frac{\log((y - y_1)/2)}{\log 3} \leq \frac{\log(3B/2)}{\log 3}, \tag{2.15}$$

therefore,

$$3^{x_1} < \frac{3B}{2} < 2B.$$

Similarly,

$$2^{y_1} \mid (3^x - 3^{x_1}) = 3^{x_1}(3^{x-x_1} - 1).$$

Analogously, if $m \geq 3$, then $2^m \mid (3^n - 1)$ if and only if $2^{m-2} \mid n$. Thus,

$$y_1 \leq 2 + \frac{\log(x - x_1)}{\log 2} < \frac{\log(4B)}{\log 2}, \tag{2.16}$$

and therefore

$$2^{y_1} \leq 4B.$$

The original equation may be rewritten in such a way that the *large parts* are on one side and the *small parts* are on the other, namely

$$|3^x - 2^y| = |3^{x_1} - 2^{y_1}| < 2B,$$

which in turn gives an inequality of the form

$$|1 - 3^x 2^{-y}| < \frac{2B}{2^B}.$$

Thus the linear form Λ to study is given by

$$\Lambda = x \log 3 - y \log 2.$$

If $\Lambda > 0$, then

$$e^\Lambda - 1 < \frac{2B}{2^B}.$$

If $\Lambda < 0$, assuming that $B > 10$, we find $\frac{2B}{2^B} < \frac{1}{2}$ and therefore, $|1 - e^\Lambda| < \frac{1}{2}$, which implies $e^{|\Lambda|} < 2$. In particular,

$$|\Lambda| < \frac{4B}{2^B}. \tag{2.17}$$

The last inequality holds independent of the sign of Λ. We observe that $\Lambda \neq 0$, since in the opposite case, we would get $3^x = 2^y$ which, by unique factorization, implies $x = y = 0$, a contradiction. Put

$$\alpha_1 = 2, \quad \alpha_2 = 3, \quad b_1 = y, \quad b_2 = x, \quad B = y, \quad A_1 = 1, \quad A_2 = \log 3$$

and

$$b' = \frac{y}{\log 3} + x < B\left(\frac{1}{\log 3} + 1\right) < 2B.$$

Since log 2 and log 3 are linearly independent, real and positive, Lemma 2.4.19 yields the estimate

$$\log|\Lambda| > -23.34(\max\{\log(2B) + 0.14, 21\})^2 \cdot \log 3.$$

Comparing the last inequality with (2.17), we get

$$B \log 2 - \log(4B) < 23.34 \cdot 3 \cdot (\max\{\log(2B) + 0.14, 21\})^2.$$

If the above maximum is 21, we get

$$B \log 2 - \log(4B) < 25.7 \cdot 21^2,$$

hence $B < 17000$. Otherwise, we have

$$B \log 2 - \log(4B) < 25.7(\log(2B) + 0.14)^2,$$

yielding $B < 2900$. Thus, we consider the inequality $B < 17000$. From (2.15) and (2.16), we get $x_1 \leq 9$ and $y_1 \leq 16$. Hence,

$$x - 1 \leq (y - 1)\frac{\log 2}{\log 3} < B\frac{\log 2}{\log 3} < 11000.$$

Now, we reduce this bound. Suppose that $B \geq 30$, then we get

$$3^x > 3^x - 3^{x_1} = 2^y - 2^{y_1} \geq 2^{B-1} \geq 2^{29},$$

which implies $x \geq 19$. We check that the congruence

$$3^x - 3^{x_1} - 2^{y_1} \equiv 0 \pmod{2^{30}}$$

does not hold for any triple (x, x_1, y_1) with $11 \leq x \leq 1100, 0 \leq x_1 \leq 9$, and $0 \leq y_1 \leq 16$. This gives $B \leq 29$. Since $3^{x-1} < 2^{y-1} \leq 2^{28}$, we get $x \leq 18$. Now, it is easy to show that there are no solutions beyond those in the statement of the proposition. For details see [36].

2.4.9   The Diophantine Equation $ax^n - by^n = c$

We consider

$$ax^n - by^n = c,$$

where a, b are strictly positive and x, y, n are unknowns. If for some exponent n there exists a solution (x, y) with $|y| > 1$, then

$$\Lambda = \log \left| \frac{a}{b} \right| - n \log \left| \frac{x}{y} \right| = O(|y|^{-n}).$$

In the other direction, Matveev's Theorem 2.2.14 implies

$$\log |\Lambda| \geq -c_* \log |y| \log n.$$

Comparing both estimates, we get $n < c_{**}$, where c_{**} depends only on a, b, c.
The following theorems give us explicit results.

Theorem 2.4.23 (Mignotte, [40]) *Assume that the exponential Diophantine inequality*

$$|ax^n - by^n| \leq c, \ a, b, c \in \mathbb{Z}_+, \ a \neq b$$

has a solution in positive integers x, y with $\max\{x, y\} > 1$. Then

$$n \leq \max \left\{ 3 \log(1.5|c/b|), \ 7400 \frac{\log A}{\log(1 + (\log A)/\log |a/b|)} \right\}, \ A = \max\{a, b, 3\}.$$

Bennett obtained the following definitive result for $c = \pm 1$.

Theorem 2.4.24 (Bennett, [8]) *For $n \geq 3$, the equation*

$$|ax^n - by^n| = 1, \ a, b \in \mathbb{Z}_+$$

has at most one solution in positive integers x, y.

For more examples of applications of linear forms in logarithms we refer to [25, 36] which are highly recommended for this purpose.

Acknowledgements The first author, Sanda Bujačić, would like to express her sincere gratitude to Prof. Jörn Steuding for organizing summer school *Diophantine Analysis* in Würzburg in 2014 and for inviting her to organize the course *Linear forms in logarithms*. She thanks him for his patience, kindness and the motivation he provided to bring this notes to publishing.

Besides Prof. Steuding, she would like to thank her PhD supervisor, Prof. Andrej Dujella, for his insightful comments during her PhD study, great advices in literature that was used for creating these lecture notes and his constant encouragement. She would also like to thank her co-author, Prof. Alan Filipin, for his kind assistance, guidance, help and excellent cooperation.

Last but not the least, she would like to thank her family: parents, sister and boyfriend for supporting her throughout writing, teaching and her life in general.

Both authors are supported by Croatian Science Foundation grant number 6422.

References

1. A. Baker, Linear forms in the logarithms of algebraic numbers, I. Mathematika J. Pure Appl. Math. **13**, 204–216 (1966)
2. A. Baker, Linear forms in the logarithms of algebraic numbers, II. Mathematika.J. Pure Appl. Math. **14**, 102–107 (1967)
3. A. Baker, Linear forms in the logarithms of algebraic numbers, III. Mathematika J. Pure Appl. Math. **14**, 220–228 (1967)
4. A. Baker, *Transcendental Number Theory* (Cambridge University Press, Cambridge, 1975)
5. A. Baker, H. Davenport, The equations $3x^2 - 2 = y^2$ and $8x^2 - 7 = z^2$. Quart. J. Math. Oxford Ser. **20**(2), 129–137 (1969)
6. A. Baker, G. Wüstholz, Logarithmic forms and group varieties. J. für die Reine und Angewandte Mathematik **442**, 19–62 (1993)
7. M. Bennett, On some exponential Diophantine equations of S. S. Pillai. Canad. J. Math. **53**, 897–922 (2001)
8. M. Bennett, Rational approximation to algebraic numbers of small height: the Diophantine equation $|ax^n - by^n| = 1$. J. Reine Angew. Math. **535**, 1–49 (2001)
9. E. Borel, Contribution a l'analyse arithmétique du continu. J. Math. Pures Appl. **9**, 329–375 (1903)
10. Y. Bugeaud, M. Mignotte, S. Siksek, Classical and modular approaches to exponential Diophantine equations I. Fibonacci and Lucas perfect powers. Ann. Math. **163**(3), 969–1018 (2006)
11. E.B. Burger, R. Tubbs, *Making Transcendence Transparent: An Intuitive Approach to Classical Transcendental Number Theory* (Springer, New York, 2004)
12. J.W.S. Cassels, *An Introduction to Diophantine Approximation, Cambridge Tracts in Mathematics and Mathematical Physics*, vol. 45 (Cambridge University Press, Cambridge, 1957)
13. H. Cohen, *Number Theory, Volume I: Tools and Diophantine Equations* (Springer, Berlin, 2007)
14. H. Cohen, *Number Theory, Volume II: Analytic And Modern Tools* (Springer, Berlin, 2007)
15. A. Dujella, The problem of the extension of a parametric family of Diophantine triples. Publ. Math. Debrecen **51**, 311–322 (1997)
16. A. Dujella, On the exceptional set in the problem of Diophantus and Davenport. Appl. Fibonacci Numbers **7**, 69–76 (1998)
17. A. Dujella, A proof of the Hoggatt–Bergum conjecture. Proc. Amer. Math. Soc. **127**, 1999–2005 (1999)
18. A. Dujella, There are only finitely many Diophantine quintuples. J. Reine Angew. Math. **566**, 183–214 (2004)
19. A. Dujella, Diofantske jednadžbe, course notes, Zagreb (2006/2007)
20. A. Dujella, Diofantske aproksimacije i primjene, course notes, Zagreb (2011/2012)
21. A. Dujella, C. Fuchs, Complete solution of a problem of Diophantus and Euler. J. London Math. Soc. **71**, 33–52 (2005)
22. A. Dujella, A. Pethő, Generalization of a theorem of Baker and Davenport. Quart. J. Math. Oxford Ser. (2) **49**, 291–306 (1998)
23. A. Dujella, A. Filipin, C. Fuchs, Effective solution of the $D(-1)$- quadruple conjecture. Acta Arith. **128**, 319–338 (2007)
24. C. Elsholtz, A. Filipin, Y. Fujita, On Diophantine quintuples and $D(-1)$ -quadruples. Monatsh. Math. **175**(2), 227–239 (2014)
25. A. Filipin, Linearne forme u logaritmima i diofantska analiza, course notes, Zagreb (2010)
26. A. Filipin, Y. Fujita, M. Mignotte, The non-extendibility of some parametric families of D(-1)-triples. Q. J. Math. **63**(3), 605–621 (2012)
27. Y. Fujita, The extensibility of $D(-1)$ -triples $\{1, b, c\}$. Publ. Math. Debrecen **70**, 103–117 (2007)
28. A.O. Gelfond, *Transcendental and Algebraic Numbers*, translated by Leo F. (Dover Publications, Boron, 1960)

29. B. He, A. Togbé, On the $D(-1)$-triple $\{1, k^2 + 1, k^2 + 2k + 2\}$ and its unique $D(1)$-extension. J. Number Theory **131**, 120–137 (2011)
30. M. Hindry, J.H. Silverman, *Diophantine Geometry: An Introduction* (Springer, New York, 2000)
31. A. Hurwitz, Ueber die angenäherte Darstellung der Irrationalzahlen durch rationale Brüche (On the approximation of irrational numbers by rational numbers). Mathematische Annalen (in German) **39**(2), 279–284 (1891)
32. S. Lang, *Introduction to Diophantine Approximations* (Addison-Wesley, Reading, 1966)
33. M. Laurent, M. Mignotte, Yu. Nesterenko, Formes linéaires en deux logarithmes et déterminants d'interpolation. J. Number Theory **55**, 285–321 (1995)
34. F. Lindemann, Über die Zahl π. Mathematische Annalen **20**, 213–225 (1882)
35. V.A. Lebesgue, Sur l'impossbilité en nombres entiers de l'équation $x^m = y^2 + 1$. Nouv. Ann. Math. **9**, 178–181 (1850)
36. F. Luca, *Diophantine Equations*, lecture notes for Winter School on Explicit Methods in Number Theory (Debrecen, Hungary, 2009)
37. K. Mahler, Zur approximation der exponentialfunktion und des logarithmus, I, II. J. reine angew. Math. **166**, 118–136, 136–150 (1932)
38. K. Mahler, Arithmetische Eigenschaften einer Klasse von Dezimalbruchen. Proc. Kon. Nederlansche Akad. Wetensch. **40**, 421–428 (1937)
39. E. M. Matveev, An explicit lower bound for a homogeneous rational linear form in logarithms of algebraic numbers I, II, Izvestiya: Mathematics, **62**(4), 723–772 (1998); **64**(6), 125–180 (2000)
40. M. Mignotte, A note on the equation $ax^n - by^n = c$. Acta Arith. **75**, 287–295 (1996)
41. O. Perron, *Die Lehre von den Kettenbrüchen* (Chelsea, New York, 1950)
42. S. S. Pillai, On $a^x - b^y = c$, J. Indian Math. Soc. (N.S.) (2), 119–122 (1936)
43. K.F. Roth, Rational approximations to algebraic numbers. Mathematika **2**, 1–20 (1955)
44. J.D. Sally, P.J. Sally Jr., *Roots to Research: A Vertical Development of Mathematical Problems* (American Mathematical Society, Providence, 2007)
45. W.M. Schmidt, *Diophantine Approximation*, vol. 785, Lecture Notes in Mathematics (Springer, Berlin, 1980)
46. J. Steuding, *Diophantine Analysis (Discrete Mathematics and Its Applications)* (Chapman& Hall/CRC, Taylor & Francis Group, Boca Raton, 2005)
47. A. Thue, Über Annäherungswerte algebraischer Zahlen. J. Reine und Angew. Math. **135**, 284–305 (1909)
48. T. Vahlen, Über Näherungswerthe und Kettenbrüche, J. Reine Angew. Math. (Crelle), **115**(3), 221–233 (1895)

Metric Diophantine Approximation—From Continued Fractions to Fractals

Simon Kristensen

2000 Mathematics Subject Classification 11J83 · 28A80

Introduction

How does one prove the existence of a real number with certain desired Diophantine properties without knowing a procedure to construct it? And if one also requires the digits in the decimal expansion, say, of this number to be special in some way, is the task then completely impossible? The present notes aim at introducing a number of methods for accomplishing this. Our main tools will be methods from classical Diophantine approximation, from dynamical systems and not least from measure theory. We will assume an acquaintance with basic measure and probability theory and some elementary number theory, but otherwise the notes aim at being self-contained.

The notes are structured as followed. We begin in Sect. 1 with some first and elementary observations on Diophantine approximation and recall some results on continued fractions. Here, we set the scene for the following sections and deduce some first metrical results. In Sect. 2, we relate the machinery of continued fractions to that of ergodic theory. We will use this machinery to deduce Khintchine's theorem in metric Diophantine approximation, which can be seen as a starting point for the metric theory of Diophantine approximation. In Sect. 3, we introduce several notions from fractal geometry. We will discuss Hausdorff measures and Hausdorff dimension, box counting dimension and Fourier dimension. We relate these to sets of arithmetical interest arising both from Diophantine approximation and from representations of real numbers in some integer base. In Sect. 4, we turn our attention

S. Kristensen (✉)
Department of Mathematics, Aarhus University,
Ny Munkegade 118, 8000 Aarhus C, Denmark
e-mail: sik@math.au.dk

© Springer International Publishing AG 2016
J. Steuding (ed.), *Diophantine Analysis*, Trends in Mathematics,
DOI 10.1007/978-3-319-48817-2_2

to higher dimensional problems. The underlying reason for this is two-fold. In a first instance, approximation of real numbers by real algebraic numbers is a higher dimensional problem. A full description of this is unfortunately beyond the scope of these notes, but we briefly outline some results and dwell a little on a conjecture on digit distribution for algebraic irrational numbers. We then turn to our second objective. Here, we study simultaneous and dual approximation of vectors of real numbers and their relation. We will also outline a proof of a higher dimensional variant of Khintchine's theorem. This will be used as a stepping stone for discussing some famous open problems in Diophantine approximation: the Duffin–Schaeffer conjecture and the Littlewood conjecture.

Several null sets of interest arise from the Khintchine type results described. One is the set of elements for which the simple approximation properties which may be derived from variants of the pigeon hole principle cannot be improved beyond a constant. In Sect. 5, we will give a general framework for studying the fractal structure of sets of such elements. In Sect. 6, we discuss the other interesting null sets arising from the Khintchine type theorems. We will outline methods for getting the Hausdorff dimension of these null sets, we will discuss approximation of elements in the ternary Cantor set by algebraic numbers, and finally we will give some results on Littlewood's conjecture. In this final part, ideas from continued fractions, uniform distribution theory, Hausdorff dimension and Fourier analysis come together in a nice blend.

1 Beginnings

Any course on Diophantine approximation should begin with the celebrated result of Dirichlet [18]:

Theorem 1.1 *Let $x \in \mathbb{R}$ and let N be a positive integer. There exist numbers $p \in \mathbb{Z}$ and $q \in \mathbb{N}$ with $q \leq N$ such that*

$$\left| x - \frac{p}{q} \right| < \frac{1}{qN}.$$

Proof Let $[x]$ denote the integer part of x and $\{x\}$ its fractional part, so that $x = [x] + \{x\}$. Divide the interval $[0, 1)$ into N subintervals $[k/N, (k+1)/N)$, where $k = 0, 1, \ldots, N - 1$, of length $1/N$. The $N + 1$ numbers $\{rx\}$, $r = 0, 1, \ldots, N$, fall into the interval $[0,1)$ and so two, $\{rx\}$ and $\{r'x\}$ say, must fall into the same subinterval, $[k/N, (k+1)/N)$ say. Suppose without loss of generality that $r > r'$. Then

$$\left| \{rx\} - \{r'x\} \right| = \left| rx - [rx] - r'x + [r'x] \right| = |qx - p| < \frac{1}{N},$$

where $q = r - r', p = [rx] - [r'x] \in \mathbb{Z}$ and $1 \leq q \leq N$. Dividing by q finishes the proof.

As an immediate corollary of Dirichlet's theorem, we obtain a non-uniform estimate.

Corollary 1.2 *Let $x \in \mathbb{R}$. For infinitely many pairs (p, q) with $p \in \mathbb{Z}$ and $q \in \mathbb{N}$,*

$$\left| x - \frac{p}{q} \right| < \frac{1}{q^2}. \tag{1.1}$$

Proof If $x \in \mathbb{Q}$, the result is trivial, as we do not require the rationals p/q to be on
lowest terms. Suppose now that $x \in \mathbb{R} \setminus \mathbb{Q}$. Fix some $N_1 \in \mathbb{N}$ and choose (p_1, q_1) as
in Dirichlet's theorem. By this theorem,

$$\left| x - \frac{p_1}{q_1} \right| < \frac{1}{q_1 N_1} \leq \frac{1}{q_1^2}.$$

As x is irrational, the left hand side must be non-zero. Consequently, it is possible to
choose an integer N_2 such that

$$\frac{1}{N_2} < q_1 \left| x - \frac{p_1}{q_1} \right|. \tag{1.2}$$

Taking this value for N in Dirichlet's Theorem gives a pair of points (p_2, q_2) with the
desired approximation property. Furthermore, $(p_1, q_1) \neq (p_2, q_2)$ since otherwise
(1.2) would contradict the choice of p_2, q_2. Continuing in this way, we obtain a
sequence of pairs p_n, q_n satisfying (1.1).

A first natural question in view of the corollary of Dirichlet's theorem is the
following: Can the rate of approximation on the right hand side be improved? In
general, the answer is negative due to a measure theoretical result. The following is
the easy half of Khintchine's theorem, which is our first example of a metric result: it
gives a condition for a certain set to be a null-set, so that almost all numbers will lie
in its complement. Throughout these notes, for a Borel set $E \subseteq \mathbb{R}^n$, we will denote
the Lebesgue measure of E by $|E|$. We will need the notion of a *limsup*-set. Recall
that given a sequence of sets E_n, we define the associated *limsup* set,

$$\limsup E_n = \bigcap_{k \geq 1} \bigcup_{n \geq k} E_n.$$

Theorem 1.3 *Let $\psi : \mathbb{N} \to \mathbb{R}_{\geq 0}$ be some function with $\sum_{q=1}^{\infty} q\psi(q) < \infty$. Then,*

$$\left| \left\{ x \in \mathbb{R} : \left| x - \frac{p}{q} \right| < \psi(q) \text{ for infinitely many } (p, q) \in \mathbb{Z} \times \mathbb{N} \right\} \right| = 0,$$

i.e. the set is a null-set with respect to the Lebesgue measure on the real line.

Proof Note first that the set is invariant under translation by integers. Hence, it suffices to prove that the set has Lebesgue measure 0 when intersected with the unit interval $[0, 1]$. Now, note that this set may be expressed as a *limsup*-set as follows,

$$\left\{ x \in [0, 1] : \left| x - \frac{p}{q} \right| < \psi(q) \text{ for infinitely many } (p, q) \in \mathbb{Z} \times \mathbb{N} \right\}$$
$$= \bigcap_{N \geq 1} \bigcup_{q \geq N} \bigcup_{p=0}^{q} \left(\frac{p}{q} - \psi(q), \frac{p}{q} + \psi(q) \right) \cap [0, 1].$$

In other words, for each $N \in \mathbb{N}$, the set is covered by

$$\bigcup_{q \geq N} \bigcup_{p=0}^{q} \left(\frac{p}{q} - \psi(q), \frac{p}{q} + \psi(q) \right),$$

so using σ-sub-additivity of the Lebesgue measure,

$$\left| \left\{ x \in [0, 1] : \left| x - \frac{p}{q} \right| < \psi(q) \text{ for infinitely many } (p, q) \in \mathbb{Z} \times \mathbb{N} \right\} \right|$$
$$\leq \left| \left(\bigcup_{q \geq N} \bigcup_{p=0}^{q} \left(\frac{p}{q} - \psi(q), \frac{p}{q} + \psi(q) \right) \right) \right| \leq \sum_{q \geq N} \sum_{p=0}^{q} \left| \left(\frac{p}{q} - \psi(q), \frac{p}{q} + \psi(q) \right) \right|$$
$$= 2 \sum_{q \geq N} (q + 1)\psi(q) \leq 4 \sum_{q \geq N} q\psi(q).$$

The latter is the tail of a convergent series, and so will tend to zero as N tends to infinity.

The connoisseur will recognise this as an application of the Borel–Cantelli lemma from probability theory. Again, the result raises more questions. Are the null-sets in fact empty? If not, what makes the elements of these sets so special? And how does one generate the infinitely many good approximants?

The usual strategy is to go via continued fractions, see e.g. [49]. There are many ways to get to these. We will go via an avenue inspired by dynamical systems (for reasons which will become clearer as we progress).

Let $x \in \mathbb{R}$ and define $a_0 = [x]$ and $r_0 = \{x\}$. If $r_0 = 0$, we stop. Otherwise, we see that $1/r_0 > 1$. Let $x_1 = 1/r_0$ and let $a_1 = [x_1]$ and $r_1 = \{x_1\}$. Continuing in this way, we define a (possibly finite) sequence $\{a_n\}$, where $a_0 \in \mathbb{Z}$ and $a_i \in \mathbb{N}$. We define a sequence of rational numbers $\{p_n/q_n\}$ by

$$\frac{p_n}{q_n} = a_0 + \cfrac{1}{a_1 + \cfrac{1}{a_2 + \cfrac{1}{\cdots + \frac{1}{a_n}}}} = [a_0; a_1, \ldots, a_n]. \tag{1.3}$$

We call the rationals p_n/q_n the *convergents to x* and the integers a_i the *partial quotients of x*. We assume that the procedure and that the following elementary properties are well-known. When $a_0 = 0$ so that $x \in [0, 1)$, we will omit this partial quotient and the semi-colon from the latter notation and write $x = [a_1, a_2, \dots]$.

Proposition 1.4 *The continued fraction algorithm has the following properties:*

(i) *The convergents may be calculated from the following recurrence formulae:*
Let $p_{-1} = 1$, $q_{-1} = 0$, $p_0 = a_0$ and $q_0 = 1$. For any $n \geq 1$,

$$p_n = a_n p_{n-1} + p_{n-2} \quad and \quad q_n = a_n q_{n-1} + q_{n-2}.$$

Consequently, $q_n \geq 2^{(n-1)/2}$.
(ii) *For any $n \geq 0$*
$$q_n p_{n-1} - q_{n-1} p_n = (-1)^n,$$

and for any $n \geq 1$,
$$q_n p_{n-2} - q_{n-2} p_n = (-1)^{n-1} a_n$$

(iii) *For an irrational number x, $x - p_n/q_n$ is positive if and only if n is even.*
(iv) *Any real irrational number x has an expansion as a continued fraction. The sequence of convergents of x converges to x, with the even (resp. odd) order convergents forming a strictly increasing (resp. decreasing) sequence. This expansion is unique, and we write $x = [a_0; a_1, \dots]$.*
(v) *Given a sequence $\{a_n\}_{n=0}^{\infty}$ with $a_0 \in \mathbb{Z}$ and $a_i \in \mathbb{N}$ for $i \geq 1$, the sequence $[a_0; a_1, \dots, a_n]$ converges to a number having the sequence $\{a_n\}$ as its sequence of partial quotients.*
(vi) *The convergents satisfy*

$$\frac{1}{q_n(q_n + q_{n+1})} < \left| x - \frac{p_n}{q_n} \right| < \frac{1}{q_n q_{n+1}} < \frac{1}{q_n^2}.$$

From Proposition 1.4 it is straightforward to construct numbers for which Corollary 1.2 can be improved. Indeed, suppose that we have the a_i for $i \leq n$ given, and let $a_{n+1} = q_n$ where q_n is given by the recursion (i). By (vi), we get

$$\left| x - \frac{p_n}{q_n} \right| < \frac{1}{q_n q_{n+1}} = \frac{1}{q_n(q_n q_n + q_{n-1})} < \frac{1}{q_n^3}. \tag{1.4}$$

By (v), the sequence $\{a_n\}$ defines an irrational number for which the exponent of Corollary 1.2 can be improved to 3 by (1.4). It is easy to modify this construction to produce an uncountable set numbers approximable with any given exponent on the right hand side.

This gives a somewhat satisfactory answer to the questions posed about the null-sets arising from Theorem 1.3. The null-sets are not empty, and the special feature of the elements of the sets is the existence of large partial quotients. Of course, the term 'large partial quotients' should now be quantified, which is where the metrical theory and the use of dynamical systems kicks in. To quantify these notions, we should ask whether there is a typical behaviour of the partial quotients, which is violated for the exceptional numbers.

2 Dynamical Methods

We consider $x \in [0, 1)$ and formalise the continued fraction algorithm in the form of a self-mapping of the unit interval.

Definition 2.1 The *Gauss map* $T : [0, 1) \to [0, 1)$ is defined by

$$Tx = \begin{cases} \{1/x\} & \text{for } x \neq 0 \\ 0 & \text{for } x = 0. \end{cases}$$

In the notation of our description of the continued fraction algorithm, we note that the Gauss map of a number $x \in [0, 1)$ extracts exactly the number $1/r_1$. Applying the Gauss map a second time, we get $T^2 x = 1/r_2$ and so on. It would seem that the Gauss map is an appropriate dynamical description of the continued fractions expansion. All we need is to get the partial quotients out of the r_i. But this can easily be done by defining the axillary function

$$a(x) = \begin{cases} [1/x] & \text{for } x \neq 0 \\ \infty & \text{for } x = 0. \end{cases} \tag{2.1}$$

We now see that

$$a_n(x) = a(T^{n-1}x), \tag{2.2}$$

where $a_n(x)$ denotes the n'th partial quotient in the continued fraction expansion of x, so iterates of the Gauss map are the natural object to study.

Having established that the Gauss map encodes the behaviour of the partial quotients, it is natural to ask for the statistical behaviour of this map – especially as we are interested in typical and a-typical behaviour of the sequence of partial quotients. A tool for this is ergodic theory. We will say that a map $T : [0, 1) \to [0, 1)$ preserves the measure μ if it is a measurable map such that for any measurable set $B \subseteq [0, 1)$, $\mu(T^{-1}B) = \mu(B)$. The Birkhoff (or pointwise) ergodic theorem is the following result (see e.g. [22]).

Theorem 2.2 (The pointwise ergodic theorem) *Let $(\Omega, \mathcal{B}, \mu)$ be a probability space and let $T : \Omega \to \Omega$ be a measure preserving transformation. Let $f \in L^1(\Omega)$. Then the limit*

$$\lim_{N \to \infty} \frac{1}{N} \sum_{n=0}^{N-1} f(T^n x) = \bar{f}(x)$$

exists for almost every $x \in \Omega$ as well as in $L^1(\Omega)$. If the transformation is ergodic, i.e. $T^{-1}B = B \Rightarrow \mu(B) \in \{0, 1\}$, the function $\bar{f}$ is constant and equal to $\int f d\mu$.

We will not prove the theorem here, but we will apply it to the Gauss map. Our approach is more or less that of [11]. The statement about the map T requires it to be measure preserving, and it is more or less self-evident that the Gauss map does not preserve the Lebesgue measure. However, there is a measure, which is absolutely continuous with respect to Lebesgue measure and with which the Gauss map is ergodic. There are good reasons why this is the correct measure, although it looks slightly mysterious at first sight. For now, we will pull the measure out of a hat and continue to work with it. Later on, we will give some indication of the origins of the measure.

Definition 2.3 Let $\mathcal{B}$ be the Borel σ-algebra in $[0, 1)$. The *Gauss measure* is defined to be the function $\mu : \mathcal{B} \to [0, 1]$ defined by

$$\mu(A) = \frac{1}{\log 2} \int_A \frac{1}{1+t} dt = \frac{1}{\log 2} \int_0^1 \chi_A(t) \frac{1}{1+t} dt.$$

Theorem 2.4 *The Gauss measure is preserved under the Gauss map, i.e. for any measurable set A, we have $\mu(T^{-1}A) = \mu(A)$.*

Proof We note that it is sufficient to prove that $\mu(T^{-1}[0, y)) = \mu([0, y))$, as we can build any other set from basic set operations on these sets. If one considers the graph of the Gauss map (try drawing it), it is easy to see that

$$T^{-1}([0, y)) = \{x \in [0, 1) : 0 \le T(x) < y\} = \bigcup_{k=1}^{\infty} \left[\frac{1}{k+y}, \frac{1}{k} \right). \qquad (2.3)$$

Thus,

$$\mu\left(T^{-1}[0, y)\right) = \sum_{k=1}^{\infty} \mu\left(\left[\frac{1}{k+y}, \frac{1}{k}\right)\right) = \sum_{k=1}^{\infty} \frac{1}{\log 2} \int_{1/(k+y)}^{1/k} \frac{1}{1+x} dx$$

$$= \sum_{k=1}^{\infty} \frac{1}{\log 2} \left[\log\left(1 + \frac{1}{k}\right) - \log\left(1 + \frac{1}{k+y}\right) \right]$$

$$= \sum_{k=1}^{\infty} \frac{1}{\log 2} \log\left(\frac{k+1}{k} \cdot \frac{k+y}{k+y+1}\right).$$

This is completely incomprehensible, so we try to get to the same incomprehensible expression from the other side. Cunningly, we make an appropriate partition and get

$$\mu([0, y)) = \frac{1}{\log 2} \int_0^y \frac{1}{1+x} dx = \sum_{k=1}^{\infty} \frac{1}{\log 2} \int_{y/(k+1)}^{y/k} \frac{1}{1+x} dx$$

$$= \sum_{k=1}^{\infty} \frac{1}{\log 2} \left[\log\left(1 + \frac{y}{k}\right) - \log\left(1 + \frac{y}{k+1}\right) \right]$$

$$= \sum_{k=1}^{\infty} \frac{1}{\log 2} \log\left(\frac{k+1}{k} \cdot \frac{k+y}{k+y+1} \right).$$

Luckily, this is the same incomprehensible mess that we have before, so the proof is complete.

Note that the density of the Gauss measure with respect to the Lebesgue measure is continuous, non-negative and in fact invertible. Hence, the two measures are absolutely continuous with respect to each other, and the property of being null or full with respect to one measure automatically implies the same for the other.

We have turned the study of the typical behaviour of continued fractions into a matter of studying the measure preserving system $([0, 1), \mathcal{B}, \mu, T)$, where $\mathcal{B}$ is the Borel σ-algebra, μ is the Gauss measure and T is the Gauss map. We have also seen that the pointwise ergodic theorem is a nice way of studying the almost everywhere behaviour of such maps. It would be desirable if the measure preserving system we have obtained turned out to be ergodic. It turns out that this is in fact the case. We will prove this now.

First, let us see what the Gauss map does to a continued fraction.

Proposition 2.5 *Let* $x = [a_1, a_2, \ldots] \in [0, 1)$. *Then*

$$Tx = T[a_1, a_2, \ldots] = [a_2, a_3, \ldots].$$

Proof We see that

$$T[a_1, a_2, \ldots] = T\left(\cfrac{1}{a_1 + \cfrac{1}{a_2 + \cfrac{1}{a_3 + \cdots}}} \right)$$

$$= \left\{ a_1 + \cfrac{1}{a_2 + \frac{1}{a_3 + \cdots}} \right\} = \cfrac{1}{a_2 + \frac{1}{a_3 + \cdots}} = [a_2, a_3, \ldots]$$

This is of course obvious from the construction. But in the light of the measure theoretic considerations, it does actually contain information. We define some sets to make life easier.

Definition 2.6 Let $a_1, \ldots, a_n \in \mathbb{N}$. Define the *fundamental interval* or *fundamental cylinder*

$$I_n(a_1, \ldots, a_n) = \{[a_1, \ldots, a_n, b_{n+1}, b_{n+2}, \ldots] : b_{n+i} \in \mathbb{N} \text{ for all } i \in \mathbb{N}\}.$$

Note that by Proposition 2.5, the n'th iterate under the Gauss map of any fundamental cylinder I_n is in fact $[0, 1) \setminus \mathbb{Q}$. This reflects the chaotic (or ergodic) nature of the Gauss map. Also note that the cylinders do not include the rational points. This is of little concern to us, as the rationals form a set of measure zero. It does however mean that we have to be extra careful with our bookkeeping.

In the following, let $n \in \mathbb{N}$ and $a_1, \ldots, a_n \in \mathbb{N}$ be fixed. Denote by I_n the fundamental interval $I_n(a_1, \ldots, a_n)$. We make a few preliminary observations.

Lemma 2.7 *We have $x \in I_n$ if and only if there exists $\theta_n(x) \in (0, 1) \setminus \mathbb{Q}$ such that*

$$x = \cfrac{1}{a_1 + \cfrac{1}{\ddots + \cfrac{1}{a_n + \theta_n(x)}}}.$$

Proof By definition, $x \in I_n$ if and only if $x = [a_1, \ldots, a_n, b_{n+1}, \ldots]$. Applying the Gauss map n times, we get

$$\theta_n(x) := T^n x = [b_{n+1}, b_{n+2}, \ldots].$$

But this is rational if and only if the sequence of partial quotients b_{n+i} terminates.

The above lemma defines a function on the irrational points in the unit interval $\theta_n : [0, 1) \setminus \mathbb{Q} \to [0, 1) \setminus \mathbb{Q}$.

Lemma 2.8 *Let $u, v \in [0, 1) \setminus \mathbb{Q}, u \le v$. Then*

$$\left| I_n \cap T^{-n}[u, v] \right| = \left| \theta_n^{-1}(v) - \theta_n^{-1}(u) \right|.$$

Proof We see that

$$I_n \cap T^{-n}[u, v] = \{x \in [0, 1) \setminus \mathbb{Q} : x = [a_1, \ldots, a_n; \theta]\}$$

where $\theta \in [u, v]$ and $[a_1, \ldots, a_n; \theta]$ denotes the continued fraction

$$x = \cfrac{1}{a_1 + \cfrac{1}{\ddots + \cfrac{1}{a_n + \theta}}}. \tag{2.4}$$

That is, for any $x \in I_n \cap T^{-n}[u, v], \theta_n(x) \in [u, v]$, so

$$I_n \cap T^{-n}[u, v] \subseteq \theta_n^{-1}[u, v].$$

Furthermore, any value of θ in $[u, v]$ inserted in (2.4) will give rise to an element in $I_n \cap T^{-n}[u, v]$, so the converse inclusion holds. Finally, it is an easy exercise left to the reader to see that θ_n^{-1} is monotonic, so this proves the lemma.

It would seem a good idea to find a precise expression for $\theta_n^{-1}(x)$. This may be done from the recursive formulae for the convergents of x.

Lemma 2.9 *We have*

$$\theta_n^{-1}(x) = \frac{p_n + x p_{n-1}}{q_n + x q_{n-1}}.$$

Proof We prove this by induction in n. For $n = 1$, using Proposition 1.4

$$\frac{p_1 + x p_0}{q_1 + x q_0} = \frac{a_1 p_0 + p_{-1} + x p_0}{a_1 q_0 + q_{-1} + x q_0} = \frac{0 + 1 + x \cdot 0}{a_1 + 0 + x} = \frac{1}{a_1 + x}$$

so $\theta_1^{-1}(x) = (p_1 + x p_0)/(q_1 + x q_0)$.

Now, we consider $n + 1$. We let $y = 1/(a_{n+1} + x)$. We know that

$$\theta_{n+1}^{-1}(x) = \cfrac{1}{a_1 + \cfrac{1}{\ddots + \cfrac{1}{a_n + \cfrac{1}{a_{n+1}+x}}}} = \cfrac{1}{a_1 + \cfrac{1}{\ddots + \cfrac{1}{a_n+y}}} = \theta_n^{-1}(y).$$

By induction hypothesis and Proposition 1.4 again,

$$\theta_n^{-1}(y) = \frac{p_n + y p_{n-1}}{q_n + y q_{n-1}} = \frac{p_n + \left(\frac{1}{a_{n+1}+x}\right) p_{n-1}}{q_n + \left(\frac{1}{a_{n+1}+x}\right) q_{n-1}}$$

$$= \frac{a_{n+1} p_n + p_{n-1} + x p_n}{a_{n+1} q_n + q_{n-1} + x q_n} = \frac{p_{n+1} + x p_n}{q_{n+1} + x q_n}.$$

This completes the proof.

Note that θ_n^{-1} is continuous, so we may extend it to all of $[0, 1]$ and still have Lemma 2.8. We now introduce the so-called Vinogradov notation to make our notation less cumbersome.

Definition 2.10 For two real expressions x and y, we say that $x \ll y$ if there exists a constant $c > 0$ such that $x \leq cy$. If $x \ll y$ and $y \ll x$ we write $x \asymp y$.

Lemma 2.11 *Let $u, v \in [0, 1)$ with $u \leq v$. Then*

$$\frac{\left|T^{-n}[u, v] \cap I_n\right|}{|I_n|} \asymp |[u, v]|,$$

where the implied constants in $\asymp$ do not depend on the sequence (a_n) defining the I_n.

Proof We use Lemma 2.8 to obtain

$$\frac{\left|T^{-n}[u, v] \cap I_n\right|}{|I_n|} = \left|\frac{\theta_n^{-1}(v) - \theta_n^{-1}(u)}{\theta_n^{-1}(1) - \theta_n^{-1}(0)}\right| = \left|\frac{\frac{p_n+vp_{n-1}}{q_n+vq_{n-1}} - \frac{p_n+up_{n-1}}{q_n+uq_{n-1}}}{\frac{p_n+p_{n-1}}{q_n+q_{n-1}} - \frac{p_n}{q_n}}\right|$$

$$= (v - u)\left|\frac{q_n(q_n + q_{n-1})}{(q_n + vq_{n-1})(q_n + uq_{n-1})}\right|.$$

The last reduction requires substantial, but completely elementary calculations using Proposition 1.4.

Now, the denominators of the convergents q_n satisfy $q_{n-1}/q_n < 1$, so it is easy to see by Proposition 1.4 (i) that

$$\left|\frac{q_n(q_n + q_{n-1})}{(q_n + vq_{n-1})(q_n + uq_{n-1})}\right| \asymp 1.$$

As $|[u, v]| = v - u$, the proof is completed.

Lemma 2.12 *For every $A \in \mathcal{B}$,*

$$\frac{\mu(T^{-n}A \cap I_n)}{\mu(I_n)} \asymp \mu(A).$$

Proof As the Borel σ-algebra is generated by intervals, by Lemma 2.11 for any $A \in \mathcal{B}$,

$$\frac{\left|T^{-n}A \cap I_n\right|}{|I_n|} \asymp |A|. \tag{2.5}$$

Also, since $1/2 < 1/(1 + t) \le 1$ for $t \in [0, 1)$, we have for any $A \in \mathcal{B}$,

$$\frac{1}{2}|A| = \int_A \frac{1}{2} dt \le \int_A \frac{1}{1 + t} dt = \mu(A)$$

and

$$\mu(A) = \int_A \frac{1}{1 + t} dt \le \int_A 1 dt = |A|.$$

Hence, $\mu(A) \asymp |A|$, so the Lemma follows from (2.5).

Theorem 2.13 *The Gauss map is ergodic with respect to the Gauss measure.*

Proof Suppose that $T^{-1}A = A$ and that $\mu(A) > 0$. It suffices to prove that $\mu(A) = 1$. Any Borel set can be generated by the I_n, as these intervals are essentially disjoint with lengths tending to zero. Hence, by generating a set B by I_n's of the same level (up to an arbitrarily small error), Lemma 2.12 implies that

$$\mu(T^{-n}A \cap B) \asymp \mu(A)\mu(B)$$

for any $B \in \mathcal{B}$. On the other hand, as $T^{-1}A = A$,

$$\mu(T^{-n}A \cap B) = \mu(A \cap B)$$

Letting $B = A^c$, we see that $\mu(A \cap B) = 0$, so that $\mu(B) \asymp 0$. This clearly implies that $\mu(B) = \mu(A^c) = 0$, so that $\mu(A) = 1$.

We specify the following corollary of the Pointwise Ergodic Theorem and Theorem 2.13:

Corollary 2.14 *Let f be an integrable function on $[0, 1)$. Then*

$$\lim_{N \to \infty} \frac{1}{N} \sum_{n=0}^{N-1} f(T^n x) = \frac{1}{\log 2} \int_0^1 \frac{f(t)}{1 + t} dt$$

for almost every $x \in [0, 1)$.

Many things follow from the ergodicity of the Gauss map. For instance, almost all numbers have an unbounded sequence of partial quotients, and in fact the arithmetic mean of the partial quotients is infinite almost surely. On the other hand, the geometric mean does have a limiting value almost surely. Calculating the typical frequency of any prescribed partial quotient is an easy exercise in integration, and combining these results with the machinery of continued fractions give a unified way in which to prove many of the classical metrical results in Diophantine approximation. An example is Lévy's theorem.

Theorem 2.15 *For almost every $x \in [0, 1)$,*

$$\lim_{n \to \infty} \frac{1}{n} \log q_n(x) = \frac{\pi^2}{12 \log 2}.$$

We will not derive this theorem here, although we will be appealing to it later. Instead, let us derive a partial converse to Theorem 1.3 due to Khintchine [49].

Theorem 2.16 *Let $\psi : \mathbb{N} \to \mathbb{R}_{\geq 0}$ be some function with $\sum_{q=1}^{\infty} q\psi(q) = \infty$ and with $q\psi(q)$ non-increasing. Then,*

$$\left| \left\{ x \in [0, 1] : \left| x - \frac{p}{q} \right| < \psi(q) \text{ for infinitely many } (p, q) \in \mathbb{Z} \times \mathbb{N} \right\} \right| = 1.$$

This is the difficult half of Khintchine's theorem, which in its totality consists of Theorem 1.3 and Theorem 2.16. It should be noted that there is an additional assumption on the function ψ. This is strictly needed. We will discuss this later in these notes.

Lemma 2.17 *Let $(\alpha_n)_{n \in \mathbb{N}}$ be a sequence of positive numbers. Suppose that*

$$\sum_{n=1}^{\infty} \frac{1}{\alpha_n} = \infty.$$

Then,

$$|\{x \in [0, 1) : a_n(x) > \alpha_n \text{ for infinitely many } n \in \mathbb{N}\}| = 1.$$

Proof We fix arbitrary $a_1, \ldots, a_n \in \mathbb{N}$ and let I_n denote the fundamental interval corresponding to these partial quotients. We let $E_n = \{x \in [0, 1) : a_n(x) > \alpha_n\}$. We first prove that

$$\frac{\mu(E_{n+1} \cap I_n)}{\mu(I_n)} \gg \frac{1}{\alpha_{n+1} + 1}. \tag{2.6}$$

In fact, all we need to do is to note that

$$E_{n+1} = T^{-n} \left\{ x \in [0, 1) : x = [\hat{a}_{n+1}, \ldots] \text{ where } \hat{a}_{n+1} > \alpha_{n+1} \right\}.$$

Thus, by Lemma 2.12,

$$\frac{\mu(E_{n+1} \cap I_n)}{\mu(I_n)} \asymp \mu \left(\{ x \in [0, 1) : x = [\hat{a}_{n+1}, \ldots] \text{ where } \hat{a}_{n+1} > \alpha_{n+1} \} \right).$$

But since the Gauss measure and the Lebesgue measure are absolutely continuous with respect to each other, the above quantity is

$$\asymp \left| \{ x \in [0, 1) : x = [\hat{a}_{n+1}, \ldots] \text{ where } \hat{a}_{n+1} > \alpha_{n+1} \} \right|$$

$$= \sum_{k > \alpha_{n+1}} \left(\frac{1}{k} - \frac{1}{k+1} \right) \gg \frac{1}{\alpha_{n+1} + 1}$$

Repeatedly using (2.6), we get for some universal $C' > 0$,

$$\mu(E_m^c \cap \cdots \cap E_{m+k}^c) \leq \prod_{i=1}^{k} \left(1 - \frac{1}{C'\alpha_{m+i} + 1} \right).$$

This holds for any $m, k \in \mathbb{N}$. To see this, note that we can express the property of being in E_m^c as being in some union of disjoint fundamental intervals. We leave the formalism as an exercise for the interested reader.

Finally,

$$\mu \left(\bigcap_{i=m}^{\infty} E_{m+i}^c \right) \leq \prod_{i=1}^{\infty} \left(1 - \frac{1}{C'\alpha_{m+i} + 1} \right).$$

As $1 - x \leq e^{-x}$ whenever $0 \leq x < 1$, we have

$$\prod_{i=1}^{n}\left(1 - \frac{1}{C'\alpha_{m+i} + 1}\right) \leq \prod_{i=1}^{n} e^{-\frac{1}{C'\alpha_{m+i}+1}} = e^{-\sum_{i=1}^{n} \frac{1}{C'\alpha_{m+i}+1}}.$$

But this tends to 0, as $\sum 1/\alpha_i$ is assumed to diverge. Thus, the probability that $a_n > \alpha_n$ only occurs finitely many times is zero, which proves the lemma.

We are now ready to prove Khintchine's Theorem.

Proof (Part II (the divergence case)) We let N be a fixed integer such that $\log N > \pi^2/12 \log 2$ ($N = 4$ will do nicely). By Theorem 2.15, for all but finitely many values of n,

$$\frac{1}{n} \log q_n(x) < \log N \tag{2.7}$$

for almost all $x \in [0, 1)$.

Let $f(q) = q\psi(q)$. Define a function $\phi(n) = N^n f(N^n)$. Since $f(q) = q\psi(q)$ is non-increasing,

$$\sum_{q=N^n}^{N^{n+1}-1} f(q) \leq (N^{n+1} - N^n)f(N^n) = (N - 1)\phi(n),$$

so as $\sum f(q)$ diverges, this will also be the case for $\sum \phi(n)$. Therefore, by Lemma 2.17, for almost every $x \in [0, 1)$,

$$a_{n+1}(x) > \frac{1}{\phi(n)}$$

holds for infinitely many n.

Now, we apply our classical estimates:

$$\left| x - \frac{p_n}{q_n} \right| \leq \frac{1}{q_n(x)q_{n+1}(x)} < \frac{1}{a_{n+1}q_n(x)^2} < \frac{\phi(n)}{q_n(x)^2}$$

for infinitely many n for almost all x. But by (2.7), $q_n(x) < N^n$, and as $qf(q)$ is non-increasing, we get

$$\phi(n) = N^n f(N^n) \leq q_n(x)f(q_n(x)).$$

Hence, for infinitely many n,

$$\left| x - \frac{p_n}{q_n} \right| < \frac{q_n(x)f(q_n(x))}{q_n(x)^2} = \psi(q_n(x)).$$

This proves the theorem.

We conclude the second section with an informal discussion of a different – and somewhat more modern – view on the Gauss map. Let us 'decompose' the action of the Gauss map into new maps. To apply the map $x \mapsto \{1/x\}$, we first apply the map $x \mapsto 1/x$ and continue to apply the map $x \mapsto x - 1$ until we finally arrive in the unit interval again. These maps are both Möbius maps given by the matrices

$$\begin{pmatrix} 0 & 1 \\ 1 & 0 \end{pmatrix} \text{ and } \begin{pmatrix} 1 & -1 \\ 0 & 1 \end{pmatrix}.$$

Recall that the Möbius map associated to a matrix

$$A = \begin{pmatrix} a & b \\ c & d \end{pmatrix}$$

is given by

$$x \mapsto \frac{ax + b}{cx + d}.$$

Composition of such maps corresponds to taking products of matrices.

For convenience, we make some sign changes here and there and consider instead the matrices

$$S = \begin{pmatrix} 0 & -1 \\ 1 & 0 \end{pmatrix} \text{ and } T = \begin{pmatrix} 1 & 1 \\ 0 & 1 \end{pmatrix}.$$

Together, these matrices generate the group $SL_2(\mathbb{Z})$.

It is tempting to look for a space on which this group acts naturally, and indeed such a space exists. The hyperbolic plane is such an object. Consider the upper half plane,

$$\mathbb{H} = \{x + iy \in \mathbb{C} : y > 0\}$$

with the Riemannian metric

$$\langle v, w \rangle_z = \frac{1}{y^2} (v \cdot w) \text{ for } z = x + iy.$$

The group $SL_2(\mathbb{R})/\{\pm I\}$ acts by isometries via Möbius maps on $\mathbb{H}$. The group $SL_2(\mathbb{Z})$ forms a lattice inside this group of isometries, and so we can consider the Riemann surface $\mathcal{M} = \mathbb{H}/SL_2(\mathbb{Z})$. This is a surface with three singularities: two points where it is non-smooth and a cusp. Applying the above generators roughly corresponds to crossing the sides of the fundamental domain of the group $SL_2(\mathbb{Z})$.

Considering the boundary of the hyperbolic plane, $\mathbb{R} \cup \{\infty\}$, we easily see that all rational numbers are identified with the point at infinity under the action of $SL_2(\mathbb{Z})$. This point in turn becomes the cusp of the surface $\mathcal{M}$.

If one formalises the above discussion, one may prove that the continued fraction of a number corresponds to a geodesic on the surface $\mathcal{M}$. Formalising this is beyond the scope of these notes, but the reader is referred to the paper [46] or the monograph

[22]. The ergodicity of the Gauss map can be seen as an instance of the ergodicity of the geodesic flow on $\mathcal{M}$ (or more generally on surfaces of constant negative curvature). The geodesic flow on $\mathcal{M}$ is a flow in the unit tangent bundle of $\mathcal{M}$, which may be identified with $\mathrm{SL}_2(\mathbb{R})/\mathrm{SL}_2(\mathbb{Z})$. This can be seen as the starting point of the use of homogeneous dynamics in Diophantine approximation, an approach which has become tremendously important in recent years. It also gives an explanation for the origins of the curious density of the Gauss measure, which can be induced on the unit interval from the natural measure on $\mathcal{M}$. In a sense, the hyperbolic measure can be seen as an instance of the hat out of which we previously pulled the Gauss measure.

3   Fractal Geometry – A Crash Course

As we saw in the last section, there is a nice zero–one dichotomy as far as the Lebesgue measure of the set of real numbers with prescribed approximation properties is concerned. However, the null sets obtained in the convergence case of Khintchine's theorem are not empty. One could easily use the machinery of continued fractions to prove that a given set is uncountable, but in fact we may discriminate even more precisely between the sizes of the null sets using the notion of Hausdorff dimension.

Hausdorff dimension was introduced by Felix Hausdorff [27], building on the construction of the Lebesgue measure given by Carathéodory [15]. Carathéodory constructed the Lebesgue measure by approximating a set E by countable covers of simple sets. The simple sets would have a volume, which could be calculated by elementary means. On adding these countably many volumes, Carathéodory would obtain an upper bound on the volume of the set E. To get the Lebesgue measure, one takes the infimum over all such covers. This produces an outer measure for which the Borel sets are measurable.

Hausdorff made the simple but far-reaching observation that if one replaces the usual volume of the sets in the covers by an appropriate function of their diameter, a different measure would be obtained. The usual volume of a hypercube in $\mathbb{R}^n$ is a constant multiple of its diameter raised to the power n, with the constant depending only on n, so this is an entirely natural thing to do. It turns out that an abundance of sets supporting such a measure exist. In particular, with the added flexibility of using different functions, one can discriminate between the sizes of Lebesgue null sets.

Let us be more concrete. For a given countable cover, $\mathcal{C}$ say, of E we consider the following sum sometimes termed the *s-length* of the cover $\mathcal{C}$, given by

$$\ell^s(\mathcal{C}) := \sum_{U \in \mathcal{C}} (\mathrm{diam}\, U)^s,$$

where $\mathrm{diam}\, U = \sup\{|\mathbf{x} - \mathbf{y}| : \mathbf{x}, \mathbf{y} \in U\}$ is the diameter of U and where $s \geq 0$ is some real number. We will also consider yet another generalisation of the above, also considered by Hausdorff. A *dimension function $f : \mathbb{R}_+ \to \mathbb{R}_+$* is a continuous, monotonic function with $f(r) \to 0$ as $r \to 0$. In the above, we may replace (diam U)^s by f(diam U) for any such function to obtain the more general notion of f-length.

For clarity of exposition, in the following we consider the special case $f(r) = r^s$ corresponding to the s-length as defined above. This is associated with what is now usually called the Hausdorff dimension but also sometimes called the Hausdorff–Besicovitch dimension. The possibly infinite number $\ell^s(\mathcal{C})$ gives an indication the 's-dimensional volume' of the set E in much the same way Carathéodory would think of it. Taking yet another hint from Carathéodory, the diameter of the sets U in the cover is now restricted to be at most $\delta > 0$.

Let
$$\mathcal{H}_\delta^s(E) := \inf_{\mathcal{C}_\delta} \sum_{U \in \mathcal{C}_\delta} (\text{diam } U)^s = \inf_{\mathcal{C}_\delta} \ell^s(\mathcal{C}_\circ),$$

where the infimum is taken over all covers $\mathcal{C}_\delta$ of E by sets U with diam $U \le \delta$; such covers are called δ-covers. As δ decreases, $\mathcal{H}_\delta^s$ can only increase as there are fewer U's available, i.e. if $0 < \delta < \delta'$, then

$$\mathcal{H}_{\delta'}^s(E) \le \mathcal{H}_\delta^s(E).$$

The set function $\mathcal{H}_\delta^s$ is an outer measure on $\mathbb{R}^n$. The limit $\mathcal{H}^s$ (which can be infinite) as $\delta \to 0$, given by

$$\mathcal{H}^s(E) = \lim_{\delta \to 0} \mathcal{H}_\delta^s(E) = \sup_{\delta > 0} \mathcal{H}_\delta^s(E) \in [0, \infty], \tag{3.1}$$

is however nicer to work with, as it is a regular outer measure with respect to which the Borel sets are measurable. It is usually called the *Hausdorff s-dimensional measure*. Hausdorff 1-dimensional measure coincides with 1-dimensional Lebesgue measure and in higher dimensions, Hausdorff n-dimensional measure is comparable to n-dimensional Lebesgue measure, i.e.

$$\mathcal{H}^n(E) \asymp |E|,$$

where $|E|$ is the Lebesgue measure of E and the implied constants depend only on n and not on the set E. Thus a set of positive n-dimensional Lebesgue measure has positive Hausdorff n-measure.

As the definition depends only on the diameter of the covering sets, there is no loss of generality in restricting to considering only covers consisting of open, closed or convex sets. Additionally, the resulting measure is clearly invariant under isometries, and scaling affects the measure in a completely natural way: for any $r \ge 0$,

$$\mathcal{H}^s(rE) = r^s \mathcal{H}^s(E).$$

As a function of s, the s-dimensional Hausdorff measure of a fixed set E exhibits an interesting behaviour. For a set E, $\mathcal{H}^s(E)$ is either 0 or ∞, except for possibly one value of s. To see this, the definition of $\mathcal{H}^s_\delta(E)$ implies that there is a δ-cover $\mathcal{C}_\delta$ of E such that

$$\sum_{C \in \mathcal{C}_\delta} (\text{diam } C)^s \le \mathcal{H}^s_\delta(E) + 1 \le \mathcal{H}^s(E) + 1.$$

Suppose that $\mathcal{H}^{s_0}(E)$ is finite and $s = s_0 + \varepsilon$, $\varepsilon > 0$. Then for each member C of the cover $\mathcal{C}_\delta$, $(\text{diam } C)^{s_0+\varepsilon} \le \delta^\varepsilon (\text{diam } C)^{s_0}$, so that the sum

$$\sum_{C \in \mathcal{C}_\delta} (\text{diam } C)^{s_0+\varepsilon} \le \delta^\varepsilon \sum_{C \in \mathcal{C}_\delta} (\text{diam } C)^{s_0}.$$

Hence

$$\mathcal{H}^{s_0+\varepsilon}_\delta(E) \le \sum_{C \in \mathcal{C}_\delta} (\text{diam } C)^{s_0+\varepsilon} \le \delta^\varepsilon \sum_{C \in \mathcal{C}_\delta} (\text{diam } C)^{s_0} \le \delta^\varepsilon (\mathcal{H}^{s_0}(E) + 1),$$

and so

$$0 \le \mathcal{H}^s(E) = \mathcal{H}^{s_0+\varepsilon}(E) = \lim_{\delta \to 0} \mathcal{H}^{s_0+\varepsilon}_\delta(E) \le \lim_{\delta \to 0} \delta^\varepsilon (\mathcal{H}^{s_0}(E) + 1) = 0.$$

On the other hand suppose $\mathcal{H}^{s_0}(E) > 0$. If for any $\varepsilon > 0$, $\mathcal{H}^{s_0-\varepsilon}(E)$ were finite, then by the above $\mathcal{H}^{s_0}(E) = 0$, a contradiction, whence $\mathcal{H}^{s-\varepsilon}(E) = \infty$.

To summarise, we have obtained a set function $\mathcal{H}^s$ associating to each infinite set $E \subseteq R^n$ an exponent $s_0 \ge 0$ for which

$$\mathcal{H}^s(E) = \begin{cases} \infty, & 0 \le s < s_0, \\ 0, & s_0 < s < \infty. \end{cases}$$

The critical exponent

$$s_0 = \inf\{s \in [0, \infty): \mathcal{H}^s(E) = 0\} = \sup\{s \in [0, \infty): \mathcal{H}^s(E) = \infty\}, \qquad (3.2)$$

where the Hausdorff s-measure crashes is called the *Hausdorff dimension* of the set E and is denoted by $\dim_H E$. It is clear that if $\mathcal{H}^s(E) = 0$ then $\dim_H E \le s$; and if $\mathcal{H}^s(E) > 0$ then $\dim_H E \ge s$. However, nothing is revealed from the definition about the measure at the critical exponent, and indeed it can take any value in the interval $[0, \infty]$ with ∞ included.

The main properties of Hausdorff dimension for sets in $\mathbb{R}^n$ are:

(i) If $E \subseteq F$ then $\dim_H E \le \dim_H F$.
(ii) $\dim_H E \le n$.
(iii) If $|E| > 0$, then $\dim_H E = n$.
(iv) The dimension of a point is 0.

(v) If $\dim_H E < n$, then $|E| = 0$ (however $\dim_H E = n$ does not imply $|E| > 0$).

(vi) $\dim_H(E_1 \times E_2) \geq \dim_H E_1 + \dim_H E_2$.

(vii) $\dim_H \cup_{j=1}^{\infty} E_j = \sup\{\dim_H E_j : j \in \mathbb{N}\}$.

It easily follows from the above properties that the Hausdorff dimension of any countable set is 0 and that of any open set in $\mathbb{R}^n$ is n. The nature of the construction of Hausdorff measure ensures that the Hausdorff dimension of a set is unchanged by an invertible transformation which is bi-Lipschitz. This implies that for any set $S \subseteq \mathbb{R} \setminus \{0\}$, $\dim_H S^{-1} = \dim_H S$, where $S^{-1} = \{s^{-1} : s \in S\}$. To see this, we split up the positive real axis into intervals $(\frac{1}{n}, \frac{1}{n-1}]$ for $n \geq 2$ together with the intervals $(m, m+1]$ for $m \geq 1$. The negative real axis is similarly decomposed. On each interval, the map $s \mapsto s^{-1}$ is bi-Lipschitz, and so the statement follows by appealing to (vii) above.

Thus on the whole, Hausdorff dimension behaves as a dimension should, although that the natural formula $\dim_H(E_1 \times E_2) = \dim_H E_1 + \dim_H E_2$ does *not* always hold (it does hold for certain sets, e.g., cylinders, such as $E \times I$, where I is an interval: $\dim_H(E \times I) = \dim_H E + \dim_H I = \dim_H E + 1$ by (iii), see [24]).

It is often convenient to restrict the elements in the δ-covers of a set to simpler sets such as balls or cubes. For example, covers consisting of hypercubes

$$H = \{\mathbf{x} \in \mathbb{R}^n : |\mathbf{x} - \mathbf{a}|_\infty < \delta\},$$

where $|\mathbf{x}|_\infty = \max\{|x_j| : 1 \leq j \leq n\}$ is the *height* of $\mathbf{x} \in \mathbb{R}^n$, centred at $\mathbf{a} \in \mathbb{R}^n$ and with sides of length 2δ are used extensively. While outer measures corresponding to these more convenient restricted covers are not the same as Hausdorff measure, they are comparable and so have the same critical exponent. Thus there is no loss as far as dimension is concerned if the sets U are chosen to be balls or hypercubes.

Of course, the two measures are identical for sets with Hausdorff s-measure which is either 0 or ∞. Such sets are said to obey a '0-∞' law, this being the appropriate analogue of the more familiar '0-1' law in probability. Sets which do not satisfy a 0-∞ law, i.e. sets which satisfy

$$0 < \mathcal{H}^{\dim_H E}(E) < \infty, \tag{3.3}$$

are called *s-sets*; these occur surprisingly often and enjoy some nice properties. One example is the Cantor set which has Hausdorff s-measure 1 when $s = \log 2/\log 3$.

However it seems that s-sets are of minor interest in Diophantine approximation where the sets that arise naturally, such as the set of badly approximable numbers or the set of numbers approximable to a given order (see next section), obey a 0-∞ law. The first steps in this direction were taken by Jarník, who proved that the Hausdorff s-measure of the set of numbers rationally approximable to order v was 0 or ∞. This result turns on an idea related to density of Hausdorff measure.

Lemma 3.1 *Let E be a null set in $\mathbb{R}$ and let $s \in [0, 1]$. Suppose that there exists a constant $K > 0$ such that for any interval (a, b) and $s \in [0, 1]$,*

$$\mathcal{H}^s(E \cap (a, b)) \leq K(b - a)\mathcal{H}^s(E). \tag{3.4}$$

Then $\mathcal{H}^s(E) = 0$ or ∞.

Proof Suppose the contrary, i.e. suppose $0 < \mathcal{H}^s(E) < \infty$ and let K be as in the statement of the theorem. Since E is null, there exists a cover of E by open intervals (a_j, b_j) such that

$$\sum_j (b_j - a_j) < \frac{1}{K}.$$

By (3.4),

$$\mathcal{H}^s(E) = \mathcal{H}^s \left(\bigcup_j (a_j, b_j) \cap E \right) \leq K\mathcal{H}^s(E) \sum_j (b_j - a_j) < \mathcal{H}^s(E),$$

a contradiction.

The proof for a general outer measure is essentially the same. The sets we encounter in Diophantine approximation are generally not s-sets and some satisfy this 'quasi-independence' property. For instance, it was shown in [14], using a variant of the above lemma, that there is no dimension function such that the associated Hausdorff measure of the set of Liouville numbers (defined below) in an interval is positive and finite. Liouville numbers are those real numbers x for which we for any $v > 0$ can find a rational p/q such that

$$0 < \left| x - \frac{p}{q} \right| < \frac{1}{q^v}.$$

Note that in this case, the existence of a single rational p/q for each $v > 0$ implies the existence of infinitely many. We prove below that this set has Hausdorff dimension 0. The result of [14] shows that even with a general Hausdorff measure $\mathcal{H}^f$, we still cannot get a positive and finite measure.

Unless some general result is available, the Hausdorff dimension $\dim_H E$ of a null set E is usually determined in two steps, with the correct upward inequality $\dim_H E \leq s_0$ and downward inequality $\dim_H E \geq s_0$ being established separately.

For limsup sets, such as the one in Theorems 1.3 and 2.16, the Hausdorff measure version of the Borel–Cantelli lemma is often useful.

Lemma 3.2 *Let (E_k) be some sequence of arbitrary sets in $\mathbb{R}^n$ and let*

$$E = \{\mathbf{x} \in \mathbb{R}^n : \mathbf{x} \in E_k \text{ for infinitely many } k \in \mathbb{N}\}.$$

If for some $s > 0$,

$$\sum_{k=1}^{\infty} \mathrm{diam}(E_k)^s < \infty, \tag{3.5}$$

then $\mathcal{H}^s(E) = 0$ and $\dim_H E \le s$.

Proof From the definition, for each $N = 1, 2, \ldots$,

$$E \subseteq \bigcup_{k=N}^{\infty} E_k,$$

so that the family $\mathcal{C}^{(N)} = \{E_k : k \ge N\}$ is a cover for E. By (3.5),

$$\lim_{N \to \infty} \sum_{k=N}^{\infty} \mathrm{diam}(E_k)^s = 0.$$

Hence $\lim_{k \to \infty} \mathrm{diam}(E_k) = 0$ and therefore given $\delta > 0$, $\mathcal{C}^{(N)}$ is a δ-cover of E for N sufficiently large. But

$$\mathcal{H}^s_\delta(E) = \inf_{\mathcal{C}_\delta} \sum_{U \in \mathcal{C}_\circ} (\mathrm{diam}\, U)^s \le \ell^s(\mathcal{C}^{(N)}) = \sum_{k=N}^{\infty} \mathrm{diam}(E_k)^s \to 0$$

as $N \to \infty$. Thus $\mathcal{H}^s_\delta(E) = 0$ and by (3.1), $\mathcal{H}^s(E) = 0$, whence $\dim_H E \le s$.

This was essentially what we did in the last section to prove the easy half of Khintchine's theorem. It follows *mutatis mutandis* from that proof that the Hausdorff dimension of the set

$$\left\{ x \in \mathbb{R} : \left| x - \frac{p}{q} \right| < \psi(q) \text{ for infinitely many } (p, q) \in \mathbb{Z} \times \mathbb{N} \right\},$$

is at most s whenever $\sum_{q=1}^{\infty} q\psi(q)^s < \infty$, so that if $\psi(q) = q^{-\upsilon}$, the Hausdorff dimension would be at most $2/\upsilon$. As with Khintchine's theorem, this is sharp, but the converse inequality is more difficult to prove. We return to it in the final section.

Various methods exist for establishing lower bounds on the Hausdorff dimension of a set. Underlying most of these methods are variants of the so-called mass distribution principle. Of course, the key difficulty is the fact that to get lower bounds, we need to consider all covers rather than just exhibiting a single cover. This is due to the definition of the Hausdorff s-measure as the infimum over all covers of the s-length. The mass distribution principle is the following simple result.

Lemma 3.3 *Let μ be a finite and positive measure supported on a bounded subset E of $\mathbb{R}^n$. Suppose that for some $s \ge 0$, there are strictly positive constants c and δ such that $\mu(B) \le c\, (\mathrm{diam}\, B)^s$ for any ball B in $\mathbb{R}^n$ with $\mathrm{diam}\, B \le \delta$. Then $\mathcal{H}^s(E) \ge \mu(E)/c$. In particular, $\dim_H E \ge s$.*

Proof Let $\{B_k\}$ be a δ-cover of E by balls B_k. Then

$$0 < \mu(E) \le \mu\left(\bigcup_k B_k\right) \le \sum_k \mu(B_k) \le c \sum_k (\text{diam } B)^s.$$

Taking infima over all such covers, we see that $\mathcal{H}^s_\delta(E) \ge \mu(E)/c$, whence on letting $\delta \to 0$,

$$\mathcal{H}^s(E) \ge \frac{\mu(E)}{c} > 0.$$

Thus if E supports a probability measure μ ($\mu(E) = 1$) with $\mu(B) \ll (\text{diam } B)^s$ for all sufficiently small balls B, then $\dim_H E \ge s$.

We now briefly discuss two other notions of dimension, namely box counting dimension and Fourier dimension. Box counting dimension (see e.g. [24]) is somewhat easier to calculate from the empirical side, although it has some serious drawbacks. Given a set $E \subseteq R^n$ and a number $\delta > 0$, let $N_\delta(E)$ denote the least number of closed balls of radius δ needed to cover E. We then define the upper and lower box counting dimensions of E as

$$\underline{\dim}_B(E) = \liminf_{\delta \to 0} \frac{\log N_\delta(E)}{-\log \delta}, \quad \overline{\dim}_B(E) = \limsup_{\delta \to 0} \frac{\log N_\delta(E)}{-\log \delta}.$$

If the values agree, we call this the box counting dimension of E.

The definition is fairly flexible, and changing the counting function to any of a number of related counting functions does not change the value. The above counting function has been chosen as a similar quantity should be familiar to readers experienced with the concept of entropy.

A major drawback of box counting dimension is that it is not associated with any measure. As a consequence of this, it is easy to construct countable sets of positive box counting dimension. In fact, one easily proves that the box counting dimension is unchanged when taking the topological closure of the set in question. Hence, any dense and countable subset of $\mathbb{R}$ has box counting dimension 1. This is problematic if one would like to apply the notion to prove the existence of transcendental numbers with certain properties using metrical methods. Indeed, as the set of algebraic numbers is countable and dense, the statement that a set has positive box counting dimension does not imply that it must contain transcendental numbers. On the other hand, proving that a set has box counting dimension zero is a much stronger statement than the corresponding one for Hausdorff dimension.

We will define one more notion of dimension, namely the Fourier dimension of a set (see [38]). For a measure μ, denote by $\hat{\mu}$ its Fourier transform, i.e.

$$\hat{\mu}(x) = \int e^{-2\pi i \xi \cdot x} d\mu(\xi).$$

We are concerned with positive Radon probability measures, so we suppose that μ is a positive, regular Borel measure with $\mu(\mathbb{R}^n) = 1$. A poor man's version of the uncertainty principle would state, that if the Fourier transform of μ decays as $|x|$

increases, the support of the measure would be 'smeared out', and so be somewhat messy.

The technical definition of the Fourier dimension of a set is as follows: Let $\dim_F(E)$ be the unique number in $[0, n]$ such that for any $s \in (0, \dim_F(E))$, there is a non-zero Radon probability measure μ with $\text{supp}(\mu) \subseteq E$ and with $|\hat{\mu}(x)| \leq |x|^{-s/2}$, and such that for any $s > \dim_F(E)$, no such measure exists.

The types of dimension mentioned here are related as follows for a set $E \subseteq \mathbb{R}^n$:

$$\dim_F(E) \leq \dim_H(E) \leq \underline{\dim}_B(E) \leq \overline{\dim}_B(E). \tag{3.6}$$

Proving the last two inequalities is straightforward, but the first one is difficult, and requires Frostman's Lemma [25], a powerful converse to the mass distribution principle of Lemma 3.3.

We end this section by relating the fractal concepts discussed so far to arithmetical issues. This will motivate the key problem considered when discussing Diophantine approximation on fractal sets.

A classical theorem of Émile Borel [12] states that almost all real numbers with respect to the Lebesgue measure are normal to any integer base $b \geq 2$ (or absolutely normal). In other words, for almost all numbers x, any block of digits occurs in the base b expansion of x with the expected frequency, independently of b. In fact, with reference to the previous section, this can be deduced from the ergodicity of the maps $T_b : [0, 1) \rightarrow [0, 1)$ given by $T_b(x) = \{bx\}$ with respect to the Lebesgue measure. The ergodicity of these maps is much easier to prove than for the Gauss map, and the deduction of Borel's result is left as an exercise.

While almost all numbers are normal to any base, the only actual examples known of such numbers are artificial and very technical to even write down (see e.g. [44]). For well-known constants such at π, $\log 2$ or even $\sqrt{2}$, very little is known about their distribution of digits. It is a long-standing conjecture that algebraic irrational numbers should be absolutely normal. In view of this, it seems natural to study which Diophantine properties a number which *fails* to be normal in some way can enjoy. One could hope that this would shed light on the question of the normality (or non-normality) of algebraic numbers.

With these remarks, let us consider a specific set of non-normal numbers which is of interest. Any such set will have Lebesgue measure zero, but in order to get anywhere with our analysis, we will take a particularly structured example. Let

$$\mathcal{C} = \left\{ x \in [0, 1] : x = \sum_{i=1}^{\infty} a_i 3^{-i}, a_i \in \{0, 2\} \right\},$$

i.e. the set of numbers in $[0, 1]$ which can be expressed in base 3 without using the digit 1. Clearly, an element of $\mathcal{C}$ cannot be absolutely normal. Of course, this set is just the well-known ternary Cantor set, and just looking at it would suggest approaching the study of the set by using fractal geometry.

Proposition 3.4 *The set $\mathcal{C}$ has $\dim_H(\mathcal{C}) = \dim_B(\mathcal{C}) = \log 2/\log 3$ and $\dim_F(\mathcal{C}) = 0$.*

We will calculate the Hausdorff dimension and the box counting dimension here. The result on the Fourier dimension is due to Kahane and Salem [31] and requires more work. However, we will calculate the Fourier transform of a particular measure on $\mathcal{C}$ and show that this does not decay.

For the upper bound on the upper box counting dimension, we will consider the obvious coverings of $\mathcal{C}$ by intervals obtained by fixing the first n coordinates of the elements in $\mathcal{C}$. There are 2^n such intervals, and each has length 3^{-n}. Hence, for $3^{-n} \leq \delta \leq 3^{-n+1}$, we find that $N_\delta(\mathcal{C}) \leq 2^n$. It follows that

$$\overline{\dim}_B(\mathcal{C}) = \limsup_{\delta \to 0} \frac{\log N_\delta(\mathcal{C})}{-\log \delta} \leq \limsup_{n \to \infty} \frac{\log 2^n}{\log 3^{n-1}} = \frac{\log 2}{\log 3}.$$

If we can get the same lower bound on the Hausdorff dimension, applying (3.6) would give us the first two equalities immediately. For this, we will apply the mass distribution principle, so we will need a probability measure.

Initially, we assign the mass 1 to the unit interval. We then divide the mass equally between the two intervals [0, 1/3] and [2/3, 1], so that each has measure 1/2. We continue in this way, at step k dividing the mass of each 'parent' interval equally between the two 'children'. The process converges to a measure supported on the Cantor set (as a sequence of measures in the weak-$*$ topology, but we skip the details). The resulting measure μ is known as the Cantor measure.

To prove that the Cantor measure is good for applying the mass distribution principle, we need an upper estimate on the measure of an interval. Let I be an interval of length <1. Pick an integer $n \geq 0$ such that $3^{-(n+1)} \leq \operatorname{diam}(I) < 3^{-n}$. In the n'th step of the Cantor construction, the minimum gap size is 3^{-n}. Hence, the interval can intersect at most one of the level n intervals, and so, setting $s = \log 2/\log 3$,

$$\mu(I) \leq 2^{-n} = 3^{-ns} \leq 3^s \operatorname{diam}(I)^s = 2\operatorname{diam}(I)^s.$$

From the mass distribution principle of Lemma 3.3, it immediately follows that $\mathcal{H}^s(\mathcal{C}) \geq 1/2$, whence $\dim_H \mathcal{C} \geq s = \log 2/\log 3$. This completes the proof of the first part of the proposition.

To finish, we will calculate the Fourier transform of the specific Cantor measure μ constructed above. Weak-$*$ convergence of the auxiliary measures imply that for any continuous function f on [0, 1],

$$\int_0^1 f(x)d\mu(x) = \lim_{n \to \infty} 2^{-n} \sum_{a_1, \ldots, a_n \in \{0,2\}} f(a_1 3^{-1} + \cdots + a_n 3^{-n}),$$

so in order to find the Fourier transform of the measure, we need to evaluate the above expression for the function $f_t(x) = e^{-2\pi i t x}$. On inserting, we find that

$$\hat{\mu}(t) = \lim_{n \to \infty} 2^{-n} \sum_{a_1, \ldots, a_n \in \{0,2\}} e^{-2\pi i t(a_1 3^{-1} + \cdots + a_n 3^{-n})}. \tag{3.7}$$

Recalling Euler's formula for the cosine function,

$$\cos(\theta) = \frac{e^{i\theta} + e^{-i\theta}}{2},$$

we find that

$$\prod_{k=1}^{n} \cos(2\pi 3^{-k}t) = \prod_{k=1}^{n} \frac{e^{2\pi i 3^{-k}t} + e^{-2\pi i 3^{-k}t}}{2}$$

$$= 2^{-n} \prod_{k=1}^{n} e^{2\pi i 3^{-k}t} \prod_{k=1}^{n} \left(e^{-2\pi i 3^{-k}2t} + 1 \right).$$

Taking absolute values, the first product becomes of absolute value 1. Expanding the latter product, whatever remains becomes a partial sum from (3.7), so that letting n tend to infinity,

$$|\hat{\mu}(t)| = \left| \prod_{k=1}^{n} \cos(2\pi 3^{-k}t) \right|.$$

It should be clear that this does not decay polynomially with t, so certainly the Cantor measure is no good if we were to believe that the Cantor set had positive Fourier dimension. Of course, this is not the case anyway.

We conclude this chapter with some remarks connecting dynamics, fractals, measures, Diophantine approximation and numeration systems. This requires a bit of functional analysis. We refer the reader to [41] for an excellent textbook on the topic.

As we have seen, to both Diophantine properties and to base b expansions, we may associate dynamical systems on the unit interval. In the former case, this was the Gauss map, and in the latter the base b map. Both of these are ergodic with respect to measures which are absolutely continuous with respect to the Lebesgue measure, so almost all numbers are typical with respect to both of these measures. Our main problem is to take an atypical property from one and prove that this forces the other to be typical.

In terms of measures, invariant sets such as the ternary Cantor set give rise to other preserved measures than the Lebesgue measure (and similarly for sets invariant under the Gauss map). A quick-and-dirty way of constructing such measures is to take a point, look at its backward orbit and take a weak limit of averages of point measures along the orbit. By the Riesz representation theorem, these measures correspond to linear functionals in the unit ball of $C([0, 1])^*$, the dual space to the continuous functions on the interval with the topology of uniform convergence. The Banach-Alaoglu theorem ensures the existence of a limit point, which again by Riesz corresponds to a measure with its support on the orbit closure of the initial point. In this way, one may construct many invariant measures for a continuous transformation of the interval.

The set of invariant measures can easily be shown to be closed (hence compact) and convex. As such, it is spanned by its extremal points, and it is again an easy exercise to prove that these are exactly the measures with respect to which the transformation is ergodic. It then follows from the pointwise ergodic theorem that such measures must be mutually singular. In the cases considered above, the space of ergodic measures has an element which is absolutely continuous with respect to Lebesgue and a whole bunch of 'fractal' measures such as the Cantor measure.

In view of this, it would seem that our problems should boil down to considering nice, convex subsets of the Banach space $C([0, 1])^*$, and subsequently study the support of elements in their intersections, when interpreted as measures by the Riesz representation theorem. However easy this may sound, it really is horrible! Indeed, one can construct a simplex in an infinite dimensional space whose extremal points are dense within it (the Poulsen simplex [43]). Evidently, the existence of such monstrous sets makes life harder for us, but it also puts the study of Diophantine approximation into a much broader context with connections all over mathematics.

4 Higher Dimensional Problems

In the last section, we mentioned a major open problem in Diophantine approximation: *Are algebraic irrational numbers absolutely normal?* In order to approach this problem, we should address the concept of approximation by algebraic numbers. This is a higher dimensional problem, and we will approach it by considering rational approximation in higher dimensional spaces. The added flexibility of having more than one variable allows us to come up with new problems as well as to state analogues of old ones in higher dimension. As it turns out, some of the unsolved problems in one dimension can be resolved in higher dimension, while some problems which naturally live in higher dimensions remain unsolved.

As our starting point, we will derive some elementary results from the geometry of numbers (see [17, 39]). Let $S \subseteq \mathbb{R}^n$ be a centrally symmetric convex set, i.e. a set S such that if $x, y \in S$ then the line segment joining x and y is fully contained in S, and so that if $x \in S$, then $-x \in S$. Of course, such sets need not be Borel, as is easily seen by taking the open unit ball in $\mathbb{R}^2$ together with a non-measurable subset of the unit sphere. However convex sets do belong to the larger class of Lebesgue measurable sets and so have a well defined volume. A first question is, how large this volume can get before we are guaranteed the existence of a point different from the origin with integer coordinates in S. Clearly, 2^n is a lower bound, as is seen by considering the cube. As it turns out, this is best possible.

Theorem 4.1 *Let S be a convex, centrally symmetric body of volume strictly greater than 2^n. Then, S contains a point from $\mathbb{Z}^n \setminus \{0\}$.*

Proof First, consider the set $S' = \frac{1}{2}S$, i.e.

$$S' = \{x \in \mathbb{R}^n : 2x \in S\}.$$

This set has volume strictly greater than 1. We divide the set up into disjoint bits

$$S'_u = \{x \in \mathbb{R}^n : u_i \leq x_i < u_i + 1\} \cap S', \quad \text{where} \quad u = (u_1, \ldots, u_n) \in \mathbb{Z}^n.$$

Now, consider the sets $S''_u = S'_u - u \subseteq [0, 1)^n$. The sum of the volumes of the S''_u is strictly greater than 1, so two of the sets must overlap. Hence, there are distinct points $x', x'' \in S'$ and distinct points $u', u'' \in \mathbb{Z}^n$, such that $x' - x'' = u' - u'' = u \in \mathbb{Z}^n \setminus \{0\}$. But by convexity and central symmetry,

$$\tfrac{1}{2}x' - \tfrac{1}{2}x'' = \tfrac{1}{2}u \in S' = \tfrac{1}{2}S,$$

so that $u \in S$.

Note that if we further assume that S is closed, the inequality of the above theorem can be weakened to $\mathrm{vol}(S) \geq 2^n$ by a simple compactness argument. We can use Theorem 4.1 to provide solutions to systems of Diophantine inequalities.

Theorem 4.2 *Let $(a_{ij}) \in \mathrm{GL}(n, \mathbb{R})$ be some invertible matrix, let $c_1, \ldots, c_n > 0$, and consider the system of inequalities*

$$\left| \sum_{j=1}^n a_{1j}x_j \right| \leq c_1$$

$$\left| \sum_{j=1}^n a_{ij}x_j \right| < c_i, \quad 2 \leq i \leq n.$$

If $c_1 \cdots c_n \geq |\det(a_{ij})|$, this system has a non-trivial integer solution.

Proof It is straightforward to verify that the system of inequalities define a centrally symmetric convex set of volume $2^n c_1 \cdots c_n |\det(a_{ij})|^{-1}$. Thus, if $c_1 \cdots c_n > |\det(a_{ij})|$, the theorem is immediately implied by Theorem 4.1.

To get the full theorem, we first replace c_1 by $c_1 + \epsilon$ for some arbitrary $\epsilon \in (0, 1)$. By the above argument, there is an integer solution $x^{(\epsilon)} \in \mathbb{Z}^n$ for each ϵ, and furthermore, the corresponding convex sets are all bounded by a constant independent of ϵ. Consequently, there are only finitely many possible integer solutions to the system of equations, so one must occur for all ϵ_k in some sequence with $\epsilon_k \to 0$. This is the point we are looking for.

Now, let $x_1, \ldots, x_n \in \mathbb{R}$, let $N \in \mathbb{N}$ and define the matrix

$$(a_{ij}) = \begin{pmatrix} 0 & 0 & \cdots & 0 & 1 \\ 1 & 0 & \cdots & 0 & -x_1 \\ 0 & 1 & \cdots & 0 & -x_2 \\ \vdots & \vdots & & \vdots & \vdots \\ 0 & 0 & \cdots & 1 & -x_n \end{pmatrix}.$$

Taking $c_1 = N^n$ and $c_2 = \cdots = c_{n+1} = 1/N$, the conditions of Theorem 4.2 are clearly satisfied. We have shown the following extension of Dirichlet's theorem.

Corollary 4.3 *Let $x_1, \ldots, x_n \in \mathbb{R}$, let $N \in \mathbb{N}$. There are integers $p_1, \ldots, p_n$ and q, $0 < q \leq N^n$ such that*

$$\left| x_i - \frac{p_i}{q} \right| < \frac{1}{qN}, \quad 1 \leq i \leq n.$$

Just as we did in the case of Dirichlet's theorem, we can derive a non-uniform version.

Corollary 4.4 *Let $x_1, \ldots, x_n \in \mathbb{R}$. There are infinitely many tuples $(p_1, \ldots, p_n) \in \mathbb{Z}^n$ and integers $q \in \mathbb{Z} \setminus \{0\}$, such that*

$$\left| x_i - \frac{p_i}{q} \right| < \frac{1}{q^{1+1/n}}, \quad 1 \leq i \leq n.$$

Considering instead the transpose of the above matrix,

$$(a_{ij}) = \begin{pmatrix} 0 & 1 & \cdots & 0 & 0 \\ \vdots & \vdots & & \vdots & \vdots \\ 0 & 0 & \cdots & 1 & 0 \\ 0 & 0 & \cdots & 0 & 1 \\ 1 & -x_n & \cdots & -x_2 & -x_1 \end{pmatrix}$$

with $c_1 = \cdots c_n = N$ and $c_{n+1} = N^{-n}$, we get another corollary.

Corollary 4.5 *Let $x_1, \ldots, x_n \in \mathbb{R}$, let $N \in \mathbb{N}$. There are integers p and $q_1, \ldots, q_n$, $0 < \max\{|q_i|\} \leq N$ such that*

$$|\mathbf{q} \cdot \mathbf{x} - p| < \frac{1}{N^n}.$$

Here and elsewhere, $\mathbf{x}$ denotes the vector with coordinates $(x_1, \ldots, x_n)$.

Writing, as is usual in number theory, $\| \cdot \|$ for the distance to the nearest integer (or nearest vector with integer coordinates in sup-norm in higher dimension) and letting $|\mathbf{q}| = \max\{|q_i|\}$, the L^∞-norm of the vector $\mathbf{q}$, we get the following corollary.

Corollary 4.6 *Let $\mathbf{x} \in \mathbb{R}^n$. There are infinitely many $\mathbf{q} \in \mathbb{Z}^n \setminus \{0\}$, such that*

$$\|\mathbf{q} \cdot \mathbf{x}\| < |\mathbf{q}|^{-n}.$$

Several things stand out here. One is the duality between the two forms of Diophantine approximation, the simultaneous approximation and the 'linear form' approximation. Another is a little better hidden. Let $\xi \in \mathbb{R}$ and consider the vector

$\mathbf{x} = (\xi, \xi^2, \ldots, \xi^n)$. Feeding this vector into Corollary 4.6 gives us infinitely many integer polynomials taking small values at ξ.

The curve given by $\Gamma = \{(x_1, \ldots, x_n) \in \mathbb{R}^n : x_i = \xi^i, \xi \in \mathbb{R}\}$ is known as a Veronese curve, and much of the field known as Diophantine approximation on manifolds has its genesis in an attempt to understand the Diophantine properties of points on these curves and so the approximation properties of real numbers by algebraic numbers. This is a natural extension of the usual approximation by rational numbers, which is the case $n = 1$. Indeed, here a linear form has just one variable, so that one considers the quantity $\|q\xi\|$, which on dividing by q gives a rational approximation to the real number ξ.

For completeness, we should mention the two alternative ways of studying algebraic approximation (the book of Bugeaud [13] is an excellent resource). For this purpose, we introduce a little notation. Let $\mathbb{A}_n$ denote the set of real, algebraic numbers of degree at most n. For an integer polynomial P, let $H(P)$ denote the naive height of P, i.e. the maximum among the absolute values of the coefficients of P. Finally, for $\alpha \in \mathbb{A}_n$, let $H(\alpha)$ denote the height of the minimal integer polynomial of α. With these definitions, we introduce two families of Diophantine exponents,

$$w_n(\xi) = \sup\{w > 0 : 0 < |P(\xi)| < H(P)^{-w} \text{ for infinitely many}$$
$$P \in \mathbb{Z}[X], \deg P \le n\},$$

and

$$w_n^*(\xi) = \sup\{w > 0 : 0 < |\xi - \alpha| < H(\alpha)^{-w-1} \text{ for infinitely many } \alpha \in \mathbb{A}_n\}.$$

The two exponents were introduced in order to classify the transcendental numbers, a topic which we will not discuss in these notes. The first should be compared with Corollary 4.6, which almost immediately tells us that unless ξ is algebraic, $w_n(\xi) \ge n$ for all $\xi \in \mathbb{R}$. Indeed, we just apply the corollary directly to the vector $(\xi, \xi^2, \ldots, \xi^n)$. The only additional thing to take care of is the fact that $|\mathbf{q}|$ is not equal to $H(P)$, as the latter takes the constant term of P into account where the former does not. However, with ξ restricted to some bounded subset of $\mathbb{R}$, the two are comparable, and the resulting difference in definitions is absorbed by the *supremum* in the definition of $w_n(\xi)$. The two exponents are related, which is to be expected: if a polynomial takes a small value at ξ, it is not too unlikely that there is a root nearby, and conversely if α is an algebraic number close to ξ, then the minimal polynomial of α probably takes a small value at ξ.

We summarise some relations due to Wirsing [48] between the exponents in the following proposition.

Proposition 4.7 *For any n and any ξ, $w_n(\xi) \ge w_n^*(\xi)$. Furthermore, if ξ is not algebraic of degree at most n, the following inequalities hold:*

$$w_n^*(\xi) \geq w_n(\xi) - n + 1$$
$$w_n^*(\xi) \geq \frac{w_n(\xi) + 1}{2}$$
$$w_n^*(\xi) \geq \frac{w_n(\xi)}{w_n(\xi) - n + 1} \tag{4.1}$$
$$w_n^*(\xi) \geq \frac{n}{4} + \frac{\sqrt{n^2 + 16n - 8}}{4}.$$

The two exponents need not be the same.

We will not prove the proposition here. However, there is an interesting point to be made. The relation between the exponents takes us into the world of transference theorems, which underlies the duality between simultaneous and linear forms approximation. We give a very general transference principle, from which many others can be derived (see [16]).

Theorem 4.8 *Consider two systems of l linearly independent linear forms, $(f_k(\mathbf{z}))$ and $(g_k(\mathbf{w}))$, all in l variables. Let $d = |\det(g_k)|$. Suppose the function*

$$\Phi(\mathbf{z}, \mathbf{w}) = \sum_k f_k(\mathbf{z}) g_k(\mathbf{w}),$$

has integer coefficients in all products of variables $z_i w_j$. If the system of inequalities

$$\max |f_k(\mathbf{z})| \leq \lambda$$

can be solved with $\mathbf{z} \in \mathbb{Z}^l \setminus \{0\}$, then so can the system of inequalities

$$\max |g_k(\mathbf{w})| \leq (l - 1)(\lambda d)^{1/(l-1)}. \tag{4.2}$$

Proof As the forms (f_k) are linearly independent, the associated homogeneous system of equations only has the zero solution. Hence, for any solution to the first system, $\mathbf{z} \in \mathbb{Z}^n \setminus \{0\}$,

$$0 < \max |f_k(\mathbf{z})| \leq \lambda.$$

Since the right hand side of (4.2) decreases with λ, we may suppose that the last inequality above is actually an equality. Finally, we can permute the forms in order to make sure that the maximum is attained for the last form and change signs to remove the absolute value. In other words, we suppose without loss of generality that $\mathbf{z} \in \mathbb{Z}^l \setminus \{0\}$ is a solution to the initial system of inequalities with

$$\max |f_k(\mathbf{z})| = f_l(\mathbf{z}) = \lambda.$$

Filling these numbers into the expression for Φ, we get a linear form in the variables $\mathbf{w}$, which together with the first $l - 1$ forms of the system $(g_k(\mathbf{w}))$ forms a system of linear forms. We may calculate its determinant, which turns out to be $f_l(\mathbf{z})d = \lambda d$. Using Theorem 4.2, the system

$$|\Phi(\mathbf{z}, \mathbf{w})| < 1, \quad |g_k(\mathbf{w})| < (\lambda d)^{1/(l-1)}, \ 1 \le k \le l-1,$$

has a non-zero integer solution $\mathbf{w}$. This certainly gives us the first $l-1$ inequalities of (4.2).

To get the final inequality, $|g_l(\mathbf{w})| < (l-1)(\lambda d)^{1/(l-i)}$, note that $\Phi(\mathbf{z}, \mathbf{w})$ is an integer by assumption, and so must be $= 0$. Hence,

$$|\lambda g_l(\mathbf{w})| = |f_l(\mathbf{z})g_l(\mathbf{w})| = \left| -\sum_{k=1}^{l-1} f_k(\mathbf{z})g_k(\mathbf{w}) \right| \le \lambda(l-1)(\lambda d)^{1/(l-1)},$$

by the triangle inequality. This completes the proof.

This theorem explains why there is a relation between a system of inequalities given by a matrix and that given by its transpose, as seen in the following theorem.

Theorem 4.9 *Let (L_i) denote a system of n linear forms in m variables and let (M_j) denote the transposed system of m linear forms in n variables. Suppose that there is an integer solution $\mathbf{x} \ne 0$ to the inequalities*

$$\|L_i(\mathbf{x})\| \le C, \quad |x_j| \le X,$$

where $0 < C < 1 \le X$. Then the system

$$\|M_j(\mathbf{u})\| \le D, \quad |u_i| \le U,$$

has a non-zero integer solution, where

$$D = (l-1)X^{(1-n)/(l-1)}C^{n/(l-1)}, \quad U = (l-1)X^{m/(l-1)}C^{(1-m)/(l-1)}, \ l = m+n.$$

Proof We introduce new variables $\mathbf{y} = (y_1, \dots, y_n)$ and $\mathbf{v} = (v_1, \dots, v_m)$ to capture the nearest integers in the systems. Hence, we define two new systems of l linear forms in l variables

$$f_k(\mathbf{x}, \mathbf{y}) = \begin{cases} C^{-1}(L_k(\mathbf{x}) + y_k) & 1 \le k \le n \\ X^{-1}x_{k-n} & n < k \le l \end{cases}$$

and

$$g_k(\mathbf{u}, \mathbf{v}) = \begin{cases} Cu_k & 1 \le k \le n \\ X(-M_{k-n}(\mathbf{u}) + v_{k-n}) & nk << l \end{cases}.$$

It is easily checked that the conditions of Theorem 4.8 hold true with $d = C^n X^m$. Applying this theorem gives a non-zero integer solution $((u), (v))$.

It remains for us to check that $\mathbf{u} \neq 0$, but this is easy: If $D < 1$ and $\mathbf{u} = 0$, the inequalities resulting from Theorem 4.8 would force $\mathbf{v} = 0$, which contradicts the initial conclusion. Hence, $\mathbf{u} \neq 0$ as required or $D \geq 1$, in which case it is trivial to solve the inequalities.

At this point, let us define some sets which will be of great importance in the next section. The set of badly approximable numbers is the set

$$\text{Bad} = \left\{ x \in \mathbb{R} : \text{For some } C(x) > 0, \left| x - \frac{p}{q} \right| \geq \frac{C(x)}{q^2} \text{ for all } \frac{p}{q} \in \mathbb{Q} \right\}.$$

From Khintchine's theorem, this set is Lebesgue null. From the theory of continued fractions, it is also the set of numbers with bounded partial quotients, so the same conclusion follows immediately from the ergodicity of the Gauss map.

Similarly to the one dimensional case, one can define the sets (also denoted Bad by abuse of notation),

$$\text{Bad} = \left\{ \mathbf{x} \in \mathbb{R}^n : \text{for some } C(\mathbf{x}) > 0, \|q\mathbf{x}\| \geq \frac{C(\mathbf{x})}{q^{1/n}} \text{ for all } q \in \mathbb{Z} \setminus \{0\} \right\}.$$

Or the corresponding linear forms version,

$$\text{Bad}^* = \left\{ \mathbf{x} \in \mathbb{R}^n : \text{for some } C(\mathbf{x}) > 0, \|\mathbf{q} \cdot \mathbf{x}\| \geq \frac{C(\mathbf{x})}{|\mathbf{q}|^n} \text{ for all } \mathbf{q} \in \mathbb{Z}^n \setminus \{0\} \right\}.$$

Corollary 4.10 *The sets* Bad *and* Bad* *are the same.*

Proof The proof is just an application of Theorem 4.9. For $\mathbf{x} \in \mathbb{R}^n$, define the linear form $M(\mathbf{t}) = \mathbf{x} \cdot \mathbf{t}$ in n variables, and let $L_i(u) = t_i u$ denote the transposed system of n linear forms in 1 variable. Clearly, $\mathbf{x} \in \text{Bad}^*$ if and only if

$$\|M(\mathbf{t}) \max\{|t_i|\}^n \geq c^*, \tag{4.3}$$

where $c^* > 0$ depends only on $\mathbf{x}$ for all non-zero integer vectors $\mathbf{t}$. Similarly, $\mathbf{x} \in \text{Bad}$ if and only if

$$\max\{\|L_i(u)\|\}^n |u| \geq c,$$

where $c > 0$ depends only on $\mathbf{x}$ for all non-zero integers u.

Suppose $\mathbf{x} \in \text{Bad}^*$. We will prove that $\mathbf{x} \in \text{Bad}$, so let $u \neq 0$ be an integer. Let $X = U$ and $C \geq \max\{\|L_i(u)\|\}$ with $0 < C < 1$, as otherwise there is nothing to prove. As $\mathbf{x} \in \text{Bad}^*$, the values of D and U of Theorem 4.9 must satisfy that $DU^n \geq c^*$, as otherwise we would have a contradiction to (4.3). However, with the relations of the theorem,

$$DU^n = nX^{(1-n)/n} C \left(nX^{1/n} \right)^n = n^{n+1} CX^{1/n},$$

so that

$$C^n X = \left(n^{-(n+1)} D U^n\right)^n \geq n^{-n(n+1)} c^{*n}.$$

This shows that if c^* is positive and exists, then $c = n^{-n(n+1)} c^{*n}$ will work as a constant to prove that $\mathbf{x} \in$ Bad. The converse is symmetrical.

Many other nice results follow from the transference technique. It is of interest to note that the above transference inequalities become equalities only at the critical exponent derived from the Dirichlet type theorems, as seen from the above corollary. This is also the case for the inequalities between the exponents of algebraic approximation, where the critical exponent for both variants is n (see (4.1)), where the inequalities become again become equalities. Transference theorems generally reveal less information away from the critical exponent.

It would be natural to conjecture that results similar to the Khintchine theorem should hold true for algebraic approximation or at least in some form for the ambient space containing a given Veronese curve. This is in fact the case, but the lack of a good analogue of continued fractions in higher dimensions is an obstacle for the methods already used to be applicable. We will give a sketch of a geometrical proof of a Khintchine type theorem for a single linear form, which is in a sense stronger than its one-dimensional analogue, as it does not assume the approximation function to be monotonic when the number of variables is at least 3. We will then discuss the monotonicity assumption in one dimension.

For the purposes of the proof, we will need a converse to the Borel–Cantelli lemma, which we state without proof. A lower bound on the measure of such a set may be found using the following lemma (see e.g. [47])

Lemma 4.11 *Let $(\Omega, \mathcal{B}, \mu)$ be a probability space and let E_n be a sequence of events. Suppose that $\sum \mu(E_n) = \infty$. Then,*

$$\mu(\limsup E_n) \geq \limsup_{Q \to \infty} \frac{\left(\sum_{n=1}^{Q} \mu(E_n)\right)^2}{\sum_{n,m=1}^{Q} \mu(E_m \cap E_n)}.$$

In particular, if the events E_n are pairwise independent, $\mu(\limsup E_n) = 1$.

We will consider the set

$$W_n(\psi) = \{\mathbf{x} \in \mathbb{R}^n : \|\mathbf{q} \cdot \mathbf{x}\| < \psi(|\mathbf{q}|) \text{ for infinitely many } \mathbf{q} \in \mathbb{Z}^n\}.$$

In this notation, the set originally considered in the first section would be

$$W_1(\psi) = \{x \in \mathbb{R} : \|qx\| < \psi(|q|) \text{ for infinitely many } q \in \mathbb{Z}\},$$

so it is a natural generalisation. In the case of $W_1(\psi)$, we showed that the set is full provided $\sum \psi(q) = \infty$ (note the change in condition due to the change in definition), and *provided the function $q\psi(q)$ was monotonic*.

Theorem 4.12 *Let* $\psi : \mathbb{N} \to \mathbb{R}_{\geq 0}$ *be some function, with* $q\psi(q)$ *monotonic if* $m = 1, 2$. *Then, the set* $W_m(\psi)$ *is full if* $\sum q^{m-1}\psi(q) = \infty$. *If the series converges, the set* $W_m(\psi)$ *is null.*

We will follow a proof given by Dodson [19]. As in the case of Khintchine's theorem, we will make some restrictions. We will consider only points in the unit square $[0, 1)^m$, which we will think of as a torus by identifying the edges. We will think of $W_m(\psi) \cap [0, 1)^m$ as a *limsup*-set, so for a fixed $\mathbf{q} \in \mathbb{Z}^m$ let

$$E_{\mathbf{q}} = \{\mathbf{x} \in \mathbb{R}^n : \|\mathbf{q} \cdot \mathbf{x}\| < \psi(|\mathbf{q}|)\},$$

so that

$$W_m(\psi) = \limsup E_{\mathbf{q}}.$$

We call the sets $E_{\mathbf{q}}$ resonant sets due to a connection with physics, which we will not explore here.

Lemma 4.13 *For each* $\mathbf{q} \in \mathbb{Z}^m$, $|E_{\mathbf{q}}| \asymp \psi(|\mathbf{q}|)$.

Sketch of proof. We sketch the argument for $m = 2$. This is a simple geometric argument. The set $E_{\mathbf{q}}$ consists of a bunch of parallel strips. Considering only the central lines of these, i.e. the solution curves to $q_1 x + q_2 y = p$ in the unit square, and matching up the sides of the square to form a torus, we obtain a closed geodesic curve on the torus. The set $E_{\mathbf{q}}$ forms a tubular neighbourhood of this geodesic. Calculating the length (roughly $|\mathbf{q}|$) and width of this strip (roughly $\psi(|\mathbf{q}|)|\mathbf{q}|^{-1}$), we arrive at the conclusion.

Lemma 4.14 *Suppose the vectors* $\mathbf{q}, \mathbf{q}' \in \mathbb{Z}^m$ *are linearly independent over* $\mathbb{R}$. *Then the corresponding resonant sets are independent in the sense of probability, i.e.* $|E_{\mathbf{q}} \cap E_{\mathbf{q}'}| = |E_{\mathbf{q}}||E_{\mathbf{q}'}|$.

Sketch of proof. Once more, we give a sketch for $m = 2$. Consider again the central geodesics of the two resonant sets. As the vectors $\mathbf{q}$ and $\mathbf{q}'$ are linearly independent, these tessellate the torus into parallelograms. There will be $|\det(\mathbf{q}, \mathbf{q}')|$ such parallelograms, where $(\mathbf{q}, \mathbf{q}')$ denotes the matrix with columns $\mathbf{q}$ and $\mathbf{q}'$.

Consider now the tubular neighbourhoods and their intersections. These will consist of a union of scaled copies of the parallelograms of the tessellation. Calculating their individual sizes as before will give the required result.

Proof of Theorem 4.12. The convergence statement is easy. The set $W_m(\psi)$ is covered by the set

$$\bigcup_{n \geq k} \bigcup_{|\mathbf{q}| \geq k} E_{\mathbf{q}},$$

so as there are roughly Q^{m-1} vectors $\mathbf{q} \in \mathbb{Z}^m$ with $|\mathbf{q}| = Q$, Lemma 4.13 immediately gives us that

$$\left| \bigcup_{n \geq k} \bigcup_{|\mathbf{q}| \geq k} E_{\mathbf{q}} \right| \ll \sum_{n \geq k} n^{m-1} \psi(n),$$

which is a tail of a convergent series.

To get the divergence half of the statement, we show that a subset has full measure. The key is to pick a sufficiently rich collection of integer vectors for which the thinned out series (the volume sum) still diverges, but for which any pair of vectors is linearly independent. Define the sets

$$S_k = \{\mathbf{q} \in \mathbb{Z}^m : \mathbf{q} \text{ is primitive, } q_m \geq 1, |\mathbf{q}| = k\}.$$

We will consider only vectors in $P = \cup S_k$.

If $\mathbf{q}, \mathbf{q}' \in P$ satisfy a linear dependence, then for some integer v, we must have $\mathbf{q} = v\mathbf{q}'$ (or the converse). It follows that v divides all the coordinates of $\mathbf{q}$, so by primitivity, $v = \pm 1$. But since the last coordinates are positive, we must have $v = 1$, whence $\mathbf{q} = \mathbf{q}'$. In other words, any pair of vectors $\mathbf{q}, \mathbf{q}' \in P$ are linearly independent, and so by Lemma 4.14 we have $|E_{\mathbf{q}} \cap E_{\mathbf{q}'}| = |E_{\mathbf{q}}||E_{\mathbf{q}'}|$.

To apply Lemma 4.11, we must ensure that the volume sum still diverges when restricted to a sum over P. In order to accomplish this, we require an asymptotic formula for the number of elements in S_k. But this is not difficult.

$$\#S_k = \sum_{\substack{|\mathbf{q}|=k, q_m \geq 1 \\ (q_1, \ldots, q_m)=1}} 1 = \sum_{\substack{|\mathbf{q}|=k, q_m \geq 1 \\ (q_1, \ldots, q_m)=h}} \sum_{d|h} \mu(d) = \sum_{d|k} \mu(d) \sum_{\substack{|\mathbf{r}|=k/d \\ r_m \geq 1}} 1$$

$$= 2^{m-2}(2m-1) \sum_{d|k} \mu(d) \left(\frac{k}{d}\right)^{m-1} + \text{error term}.$$

Here, μ denotes the Möbius function, and we have used the classical fact that $\sum_{d|n} \mu(d)$ is equal to one for $n = 1$ and equal to zero otherwise. For $m = 2$, the main term is $= 3\phi(k)$, where ϕ denotes the Euler ϕ-function. For $m \geq 3$, we have

$$\sum_{d|k} \mu(d) \left(\frac{1}{d}\right)^{m-1} = \prod_{p|k} \left(1 - \frac{1}{p^{m-1}}\right),$$

which lies between $\zeta(m-1)^{-1}$ and 1, where ζ denotes the Riemann ζ-function.

The upshot is that for $m \geq 3$, S_k contains a constant times k^{m-1} elements, and the divergence of the original series implies the divergence of the restricted series without further work. For $m = 2$ we need to average out the irregularities of the Euler function, but using the classical estimate $\sum_{n \leq N} \phi(n) = \frac{3}{\pi^2} N^2 + O(N \log N)$, we may apply Cauchy condensation over 2-adic blocks to get the divergence of the new series. This however requires the monotonicity of the function.

Quite a few remarks should be made at this point. Firstly, the result is valid even more generally than the one stated here. The full Khintchine–Groshev theorem concerns systems of linear forms, and states the set of matrices

$$W_{m,n}(\psi) = \{A \in \mathrm{Mat}_{m \times n}(\mathbb{R}) : \|\mathbf{q}A\| < \psi(|\mathbf{q}|) \text{ for infinitely many } \mathbf{q} \in \mathbb{Z}^m\},$$

is null or full according to the convergence or divergence of the series $\sum q^{m-1}\psi(q)^n$.

Secondly, the divergence assumption is not needed except for in one case. Namely, in the case $m = n = 1$, an explicit counterexample to the classical Khintchine theorem without assumption of monotonicity can be given. However, conjectures do exists, which tell us what to expect. It is natural in this context to impose the restriction that the approximating rationals should be on lowest terms. The Duffin–Schaeffer conjecture [21] states that the set

$$\left\{ x \in \mathbb{R} : \left| x - \frac{p}{q} \right| < \psi(q) \text{ for infinitely many coprime } (p, q) \in \mathbb{Z} \times \mathbb{N} \right\},$$

should be null or full according to the convergence or divergence of the series $\sum \phi(q)\psi(q)$.

The convergence half is easy, and in the case of an appropriately monotonic approximation function, the result follows immediately from condensation and Khintchine's theorem. The difficulty is in getting the result for non-monotonic error functions. However, it is known that the set must be either null or full. It is hence tempting to try to apply Lemma 4.11 to get positive measure and proceed to deduce full measure from this law. However, controlling the intersections appears to be beyond the reach of current methods. What is clear is that it is hopeless to control the individual intersections, and the entire sum must be considered at least in very long blocks at a time.

Thirdly, as in the case when the above results give a null set, it is natural to ask what the Hausdorff dimension should be. As in the case of a single number, it is easy to get an upper bound, at least in the case $\psi(q) = q^{-v}$. Just applying a covering argument in the spirit of the convergence case, one finds that in this case, $\dim_H(W_{m,n}(q \mapsto q^{-v})) \le (m-1)n + (m+n)/(1+v)$. The initial integer comes from the hyperplanes central to the resonant sets, with the fraction at the end being the really interesting component. In fact, this is sharp, and we will return to these types of estimates in the final section of the notes.

Fourthly, we did not answer the question we originally asked. Namely, we were interested in points on the Veronese curves and in a final instance in points in a fractal subset of the Veronese curve. For the full Veronese curves, these things can be done, but with considerably more difficulty. Even the convergence case was not settled before 1989 by Bernik [9] with the divergence case taking another 10 years before being settled by Beresnevich [4]. By contrast, the above results about the ambient space date back to 1938.

As a final remark on the Khintchine–Groshev theorem, in the case when one considers simultaneous approximation or more than one linear form, similar questions can be asked when different rates of approximation are required in the different variables. Again, the questions can be answered under some assumptions on the approximating functions, and whether sets arising in this way are null or full depend once again on the convergence or divergence of a certain series.

We dwell a little on this last point. Consider again Corollary 4.4, this time with $n = 2$. Littlewood suggested multiplying the two inequalities instead of considering them separately, and so to consider

$$\left| x - \frac{p_1}{q} \right| \left| y - \frac{p_2}{q} \right| < \frac{1}{q^3}$$

or more concisely,

$$q \|qx\| \|qy\| < 1. \tag{4.4}$$

By Corollary 4.4, this inequality always has infinitely many solutions for any pair (x, y), but it is a little more flexible than the original one. Indeed, one approximation could be pretty bad indeed, just as long as the other one is very good, and the inequality would still hold.

From the theory of continued fractions, we know that many numbers x exist which have $q \|qx\| > C > 0$ for all q (the badly approximable numbers, or equivalently those with bounded partial quotients), and similarly we know that there are badly approximable pairs, i.e. pairs for which Corollary 4.4 cannot be improved beyond a positive constant (we will prove this and much more in the next section). Littlewood asked whether there are pairs such that (4.4) cannot be improved beyond a constant. Due to the added flexibility, he conjectured that this should not be the case, so that for any pair (x, y),

$$\liminf_{q \to \infty} q \|qx\| \|qy\| = 0, \tag{4.5}$$

where the liminf is taken over positive integers q.

Equation (4.5) is the Littlewood conjecture. Littlewood apparently did not think that it should be too difficult, and set it as an exercise to his students in the thirties. To date, it is an important unsolved problem in Diophantine approximation. This is a case, where many of the known results are metric. Probably the most famous among them is the result of Einsiedler, Katok and Lindenstrauss [23], which states that the set of exceptions (x, y) to (4.5) must lie in a countable union of sets of box counting dimension zero. From the elementary properties of Hausdorff dimension together with (3.6), it follows that both the Hausdorff dimension and the Fourier dimension are also equal to zero.

In fact, their approach follows the approach to continued fractions via the geodesic flow outlined in the first section. There is no good analogue of continued fractions in higher dimension, but an analogue of the geodesic flow on $SL_2(\mathbb{R}) / SL_2(\mathbb{Z})$ is certainly constructible. In the classical, one-dimensional case, the geodesic flow is given by the action of the diagonal subgroup of $SL_2(\mathbb{R})$, and badly approximable numbers correspond to geodesics which remain in a compact subset of the space $SL_2(\mathbb{R}) / SL_2(\mathbb{Z})$. The approach of Einsiedler, Katok and Lindenstrauss works instead with the diagonal subgroup of $SL_3(\mathbb{R})$, acting on $SL_3(\mathbb{R}) / SL_3(\mathbb{Z})$.

This action is a two-parameter flow and its dynamics is very complicated. Nonetheless, the three authors manage to prove many things about the simplex of preserved measures described in the final section, and in particular about the extremal points. They show that the only possible ergodic measures are the Haar measure and measures with respect to which every one-parameter subgroup of the diagonal group acts with zero entropy. Readers acquainted with the notion of entropy should be able to see how this will have an impact on the box counting dimension of the support of the measure.

The relation between the flow and the Littlewood conjecture is a little technical, but briefly the pair (x, y) satisfies the Littlewood conjecture if and only if the orbit of the point

$$\begin{pmatrix} 1 & 0 & 0 \\ x & 1 & 0 \\ y & 0 & 1 \end{pmatrix} SL_3(\mathbb{Z})$$

is unbounded under the action of the semigroup

$$A^+ = \left\{ \begin{pmatrix} e^{-s-t} & 0 & 0 \\ 0 & e^s & 0 \\ 0 & 0 & e^t \end{pmatrix} : s, t \in \mathbb{R}_+ \right\}.$$

Hence, the set of exceptions can be embedded into a set of points in $SL_3(\mathbb{R})/ SL_3(\mathbb{Z})$ with bounded A^+-orbits, and the result can be deduced from the dynamical statement. It is very impressive work. We will say something non-trivial but somewhat easier in the final section of these notes.

5 Badly Approximable Elements

In this section, we will discuss badly approximable numbers, and in fact do so in higher dimensions. A consequence of our main result is Jarník's theorem [29]: the set of badly approximable numbers has maximal Hausdorff dimension. However we will prove much more, including the result that badly approximable numbers form a set of maximal dimension inside the Cantor set. The latter relates digital properties with Diophantine properties, and although it does not resolve the question of absolute normality of algebraic irrational numbers, it does provide information on which Diophantine properties a number failing spectacularly at being normal to some base can have.

The work presented in this section originated in an unfortunately failed attempt to resolve the Schmidt conjecture with Thorn and Velani [36]. We did solve other problems in the process, though. In order to present this, we need some new sets. For $i, j \geq 0$ with $i + j = 1$, denote by $\mathrm{Bad}(i, j)$ the set of (i, j)–badly approximable pairs $(x_1, x_2) \in \mathbb{R}^2$; that is $(x_1, x_2) \in \mathrm{Bad}(i, j)$ if there exists a positive constant $c(x_1, x_2)$ such that for all $q \in \mathbb{N}$

$$\max\{\|qx_1\|^{1/i}, \|qx_2\|^{1/j}\} > c(x_1, x_2) \, q^{-1}.$$

In the case $i = j = 1/2$, the set is simply the standard set of badly approximable pairs or equivalently as we saw in the last section the set of badly approximable linear forms in two variables. If $i = 0$ we identify the set $\mathrm{Bad}(0, 1)$ with $\mathbb{R} \times \mathrm{Bad}$ where Bad is the set of badly approximable numbers. That is, $\mathrm{Bad}(0, 1)$ consists of pairs (x_1, x_2) with $x_1 \in \mathbb{R}$ and $x_2 \in \mathrm{Bad}$. The roles of x_1 and x_2 are reversed if $j = 0$. In full generality, Schmidt's conjecture states that $\mathrm{Bad}(i, j) \cap \mathrm{Bad}(i', j') \neq \emptyset$. It is a simple exercise to show that if Schmidt's conjecture is false for some pairs (i, j) and (i', j') then Littlewood's conjecture in simultaneous Diophantine approximation is true.

The Schmidt conjecture was recently settled in the affirmative by Badziahin, Pollington and Velani [3], who established a stronger version. An [1] subsequently proved an even stronger result, which we remark on towards the end of this section.

We will set up a scary generalisation of the sets $\mathrm{Bad}(i, j)$. For the purposes of these notes, we will consider general metric spaces. The examples to keep in mind are nice fractal subsets of Euclidean space, such as the Cantor set or the Sierpiński gasket. Let (X, d) be the product space of t metric spaces (X_i, d_i) and let (Ω, d) be a compact subspace of X which contains the support of a non-atomic finite measure m.

Let $\mathcal{R} = \{R_\alpha \subseteq X : \alpha \in J\}$ be a family of subsets R_α of X indexed by an infinite, countable set J. Thus, each resonant set R_α can be split into its t components $R_{\alpha,i} \subset (X_i, d_i)$. Let $\beta : J \to \mathbb{R}_+ : \alpha \to \beta_\alpha$ be a positive function on J and assume that the number of $\alpha \in J$ with β_α bounded from above is finite. We think of these as the resonant sets similar to the central lines of the resonant neighbourhoods considered in the last section.

For each $1 \leq i \leq t$, let $\rho_i : \mathbb{R}_+ \to \mathbb{R}_+ : r \to \rho_i(r)$ be a real, positive function such that $\rho_i(r) \to 0$ as $r \to \infty$ and that ρ_i is decreasing for r large enough. Furthermore, assume that $\rho_1(r) \geq \rho_2(r) \geq \cdots \geq \rho_t(r)$ for r large – the ordering is irrelevant. Given a resonant set R_α, let

$$F_\alpha(\rho_1, \ldots, \rho_t) = \{x \in X : d_i(x_i, R_{\alpha,i}) \leq \rho_i(\beta_\alpha) \quad \text{for all } 1 \leq i \leq t\}$$

denote the 'rectangular' $(\rho_1, \ldots, \rho_t)$–neighbourhood of R_α. For a real number $c > 0$, we will define the scaled rectangle,

$$cF_\alpha(\rho_1, \ldots, \rho_t) = \{x \in X : d_i(x_i, R_{\alpha,i}) \leq c\rho_i(\beta_\alpha) \quad \text{for all } 1 \leq i \leq t\},$$

and similarly for other rectangular regions throughout this section. Consider the set

$$\mathrm{Bad}^*(\mathcal{R}, \beta, \rho_1, \ldots, \rho_t)$$
$$= \{x \in \Omega : \exists\, c(x) > 0 \text{ s.t. } x \notin c(x) F_\alpha(\rho_1, \ldots, \rho_t) \text{ for all } \alpha \in J\}.$$

Thus, $x \in \mathrm{Bad}^*(\mathcal{R}, \beta, \rho_1, \ldots, \rho_t)$ if there exists a constant $c(x) > 0$ such that for all $\alpha \in J$,

$$d_i(x_i, R_{\alpha,i}) \geq c(k) \rho_i(\beta_\alpha) \quad \text{for some } 1 \leq i \leq t.$$

We wish to find a suitably general framework which gives a lower bound for the Hausdorff dimension of $\mathrm{Bad}^*(\mathcal{R}, \beta, \rho_1, \ldots, \rho_t)$. Without loss of generality we shall assume that $\sup_{\alpha \in J} \rho_i(\beta_\alpha)$ is finite for each i – otherwise $\mathrm{Bad}^*(\mathcal{R}, \beta, \rho_1, \ldots, \rho_t) = \emptyset$ and there is nothing to prove.

Given $l_1, \ldots, l_t \in \mathbb{R}_+$ and $c \in \Omega$ let

$$F(c; l_1, \ldots, l_t) = \{x \in X : d_i(x_i, c_i) \leq l_i \quad \text{for all } 1 \leq i \leq t\}$$

denote the closed 'rectangle' centred at c with 'sidelengths' determined by $l_1, \ldots, l_t$. Also, for any $k > 1$ and $n \in \mathbb{N}$, let F_n denote any generic rectangle intersected with Ω, i.e. a set of the form $F(c; \rho_1(k^n), \ldots, \rho_t(k^n)) \cap \Omega$ in Ω centred at a point c in Ω. As before, $B(c, r)$ is a closed ball with centre c and radius r. The following conditions on the measure m and the functions ρ_i will play a central role in our general framework.

(A) There exists a strictly positive constant δ such that for any $c \in \Omega$

$$\liminf_{r \to 0} \frac{\log m(B(c, r))}{\log r} = \delta.$$

It is easily verified from the Mass distribution principle of Lemma 3.3 that if the measure m supported on Ω is of this type, then $\dim \Omega \geq \delta$ and so $\dim X \geq \delta$.
(B) *For $k > 1$ sufficiently large, any integer $n \geq 1$ and any $i \in \{1, \ldots, t\}$,*

$$\lambda_i^l(k) \leq \frac{\rho_i(k^n)}{\rho_i(k^{n+1})} \leq \lambda_i^u(k),$$

where λ_i^l and λ_i^u are lower and upper bounds depending only on k but not on n, such that $\lambda_i^l(k) \to \infty$ as $k \to \infty$.
(C) *There exist constants $0 < a \leq 1 \leq b$ and $l_0 > 0$ such that*

$$a \leq \frac{m(F(c; l_1, \ldots, l_t))}{m(F(c'; l_1, \ldots, l_t))} \leq b$$

for any $c, c' \in \Omega$ and any $l_1, \ldots, l_t \leq l_0$. This condition implies that rectangles of the same size centred at points of Ω have comparable m-measure.
(D) *There exist strictly positive constants D and l_0 such that*

$$\frac{m(2 F(c; l_1, \ldots, l_t))}{m(F(c; l_1, \ldots, l_t))} \leq D$$

for any $c \in \Omega$ and any $l_1, \ldots, l_t \leq l_0$. This condition simply says that the measure m is 'doubling' with respect to rectangles.

(E) *For $k > 1$ sufficiently large and any integer $n \geq 1$*

$$\frac{m(F_n)}{m(F_{n+1})} \geq \lambda(k),$$

where λ is a function depending only on k such that $\lambda(k) \to \infty$ as $k \to \infty$.

In terms of achieving a lower bound for dim Bad$^*(\mathcal{R}, \beta, \rho_1, \ldots \rho_t)$, the above four conditions are rather natural. The following final condition is in some sense the only genuine technical condition and is not particularly restrictive.

We should state at this point that if m is a product measure of measures satisfying the decay condition that there exist strictly positive constants δ and r_0 such that for $c \in \Omega$ and $r \leq r_0$

$$a \, r^\delta \leq m(B(c, r)) \leq b r^\delta, \tag{5.1}$$

where $0 < a \leq 1 \leq b$ are constants independent of the ball, then the product measure satisfies all conditions above. This is extremely useful, and missing digit sets have this property, as do all regular Cantor sets.

Theorem 5.1 *Let (X, d) be the Cartesian product space of the metric spaces $(X_1, d_1), \ldots, (X_t, d_t)$ and let (Ω, d, m) be a compact measure subspace of X. Let the measure m and the functions ρ_i satisfy conditions (A) to (E). For $k \geq k_0 > 1$, suppose there exists some $\theta \in \mathbb{R}_+$ so that for $n \geq 1$ and any rectangle F_n there exists a disjoint collection $\mathcal{C}(\theta F_n)$ of rectangles $2\theta F_{n+1}$ contained within θF_n satisfying*

$$\#\mathcal{C}(\theta F_n) \geq \kappa_1 \frac{m(\theta F_n)}{m(\theta F_{n+1})} \tag{5.2}$$

and

$$\# \left\{ 2\theta F_{n+1} \subset \mathcal{C}(\theta F_n) : \min_{\alpha \in J(n+1)} d_i(c_i, R_{\alpha,i}) \leq 2\theta \rho_i(k^{n+1}) \text{ for any } 1 \leq i \leq t \right\}$$

$$\leq \kappa_2 \frac{m(\theta F_n)}{m(\theta F_{n+1})}. \tag{5.3}$$

where $0 < \kappa_2 < \kappa_1$ are absolute constants independent of k and n. Furthermore, suppose $\dim_H(\cup_{\alpha \in J} R_\alpha) < \delta$. Then

$$\dim_H \text{Bad}^*(\mathcal{R}, \beta, \rho_1, \ldots, \rho_t) \geq \delta.$$

The statement of the theorem with all its assumptions is pretty bad, and the proof is in fact a rather dull affair. We give a short sketch. Fixing $k \geq k_0$, the conditions of the theorem give us a way to construct a Cantor type set inside Bad$^*(\mathcal{R}, \beta, \rho_1, \ldots, \rho_t)$. Namely, we begin with a rectangle θF_1. In this and any subsequent step, we take out the collection $\mathcal{C}(\theta F_n)$, which is pretty big due to (5.2). The points in the rectangles from (5.3) will have some difficulties lying in the set Bad$^*(\mathcal{R}, \beta, \rho_1, \ldots, \rho_t)$, as they are fairly close to a resonant set from $J(n+1)$. Hence, we discard them and retain a collection $\mathcal{F}_{n+1}(\theta F_n)$ of closed rectangles. The assumption (5.3) tells us that a positive proportion of the collection will remain, and we continue in this way to get a Cantor set constructed from rectangles. From the construction, this set is contained in Bad*$(\mathcal{R}, \beta, \rho_1, \ldots, \rho_t)$, and in fact the unspecified constant in the definition of the set can in all cases be chosen to be $c(k) = \min_{1 \le i \le t}(\theta/\lambda_i^u(k))$. We call the set $\mathbf{K}_{c(k)}$.

We now construct a probability measure on the Cantor set recursively. For any rectangle θF_n in $\mathcal{F}_n$ we attach a weight $\mu(\theta F_n)$ which is defined recursively as follows: for $n = 1$,

$$\mu(\theta F_1) = \frac{1}{\#\mathcal{F}_1} = 1$$

and for $n \ge 2$,

$$\mu(\theta F_n) = \frac{1}{\#\mathcal{F}_n(\theta F_{n-1})} \mu(\theta F_{n-1}) \quad (F_n \subset F_{n-1}).$$

This procedure thus defines inductively a mass on any rectangle used in the construction of $\mathbf{K}_{c(k)}$. In fact a lot more is true: μ can be further extended to all Borel subsets A of Ω to determine $\mu(A)$ so that μ constructed as above actually defines a measure supported on $\mathbf{K}_{c(k)}$. The probability measure μ constructed above is supported on $\mathbf{K}_{c(k)}$ and for any Borel subset A of Ω

$$\mu(A) = \inf \sum_{F \in \mathcal{F}} \mu(F),$$

where the infimum is taken over all coverings $\mathcal{F}$ of A by rectangles $F \in \{\mathcal{F}_n : n \ge 1\}$.

The mass distribution principle of Lemma 3.3 can then be applied to this measure to find that dim $\mathbf{K}_{c(k)} \ge \delta - 2\epsilon(k)$, where $\epsilon(k)$ tends to zero as k tends to infinity. To conclude, we let k do this, and so have constructed a subset of Bad*$(\mathcal{R}, \beta, \rho_1, \ldots, \rho_t)$ whose dimension is lower bounded by δ. All the technical assumptions occur naturally in the process of constructing the set and applying the mass distribution principle.

The remarks preceding the statement of Theorem 5.1 immediately gives us the following version, which is of more use to us.

Theorem 5.2 *For $1 \le i \le t$, let (X_i, d_i) be a metric space and (Ω_i, d_i, m_i) be a compact measure subspace of X_i where the measure m_i satisfies (5.1) with exponent δ_i. Let (X, d) be the product space of the spaces (X_i, d_i) and let (Ω, d, m) be the product measure space of the measure spaces (Ω_i, d_i, m_i). Let the functions ρ_i satisfy condition (B). For $k \ge k_0 > 1$, suppose there exists some $\theta \in \mathbb{R}_+$ so that for $n \ge 1$ and any rectangle F_n there exists a disjoint collection $C(\theta F_n)$ of rectangles $2\theta F_{n+1}$ contained within θF_n satisfying*

$$\#C(\theta F_n) \;\geq\; \kappa_1 \prod_{i=1}^{t} \left(\frac{\rho_i(k^n)}{\rho_i(k^{n+1})} \right)^{\delta_i} \tag{5.4}$$

and

$$\# \left\{ 2\theta F_{n+1} \subset C(\theta F_n) : \; \min_{\alpha \in J(n+1)} d_i(c_i, R_{\alpha,i}) \leq 2\theta \rho_i(k^{n+1}) \, \text{for any } 1 \leq i \leq t \right\}$$

$$\leq \kappa_2 \prod_{i=1}^{t} \left(\frac{\rho_i(k^n)}{\rho_i(k^{n+1})} \right)^{\delta_i} , \tag{5.5}$$

where $0 < \kappa_2 < \kappa_1$ are absolute constants independent of k and n. Furthermore, suppose $\dim_H (\cup_{\alpha \in J} R_\alpha) < \sum_{i=1}^{t} \delta_i$. *Then*

$$\dim_H \text{Bad}^*(\mathcal{R}, \beta, \rho_1, \dots \rho_t) = \sum_{i=1}^{t} \delta_i.$$

Note that while Theorem 5.1 will only give the lower bound on the Hausdorff dimension in the last equation, the upper bound is a consequence of the assumptions. Indeed, any set satisfying (5.1) will have Hausdorff dimension equal to δ, and for these particular nice sets, the Cartesian product satisfies the expected dimensional relation, so that the ambient space in the above result is of Hausdorff dimension $\sum_{i=1}^{t} \delta_i$.

The interest in Theorem 5.1 is not in its proof, but in its applications. In the original paper, in addition to the study of Bad(i, j) and similar sets, the theorem was applied to approximation of complex numbers by ratios of Gaussian integers, to approximation of p-adic numbers, to function fields over a finite field, to problems in complex dynamics and to limit sets of Kleinian groups. More applications have occurred since then.

Initially, we use it to prove Jarník's theorem. Let $I = [0, 1]$ and consider the set

$$\text{Bad}_I = \left\{ x \in [0, 1] : \left| x - \frac{p}{q} \right| > c(x)/q^2 \text{ for all rationals } \frac{p}{q} \right\}.$$

This is the classical set Bad of badly approximable numbers restricted to the unit interval. Clearly, it can be expressed in the form Bad$^*(\mathcal{R}, \beta, \rho)$ with $\rho(r) = r^{-2}$ and

$$X = \Omega = [0, 1] , \quad J = \{(p, q) \in \mathbb{N} \times \mathbb{N} \backslash \{0\} : p \leq q\} ,$$

$$\alpha = (p, q) \in J , \quad \beta_\alpha = q , \quad R_\alpha = \frac{p}{q}.$$

The metric d is of course the standard Euclidean metric; $d(x, y) := |x - y|$. Thus in this basic example, the resonant sets R_α are simply rational points p/q. With reference to our framework, let the measure m be one–dimensional Lebesgue measure on I. Thus, $\delta = 1$ and all the many conditions are easily checked.

We show that the conditions of Theorem 5.1 are satisfied for this basic example. The existence of the collection $\mathcal{C}(\theta B_n)$, where B_n is an arbitrary closed interval of length $2k^{-2n}$ follows immediately from the following simple observation. For any two distinct rationals p/q and p'/q' with $k^n \leq q, q' < k^{n+1}$ we have that

$$\left| \frac{p}{q} - \frac{p'}{q'} \right| \geq \frac{1}{qq'} > k^{-2n-2}. \tag{5.6}$$

Thus, any interval θB_n with $\theta := \frac{1}{2}k^{-2}$ contains at most one rational p/q with $k^n \leq q < k^{n+1}$. Let $\mathcal{C}(\theta B_n)$ denote the collection of intervals $2\theta B_{n+1}$ obtained by subdividing θB_n into intervals of length $2k^{-2n-4}$ starting from the left hand side of θB_n. Clearly

$$\#\mathcal{C}(\theta B_n) \geq [k^2/2] > k^2/4 = \text{r.h.s. of (5.2) with } \kappa_1 = 1/4.$$

Also, in view of the above observation, for k sufficiently large

$$\text{l.h.s. of (5.3)} \leq 1 < k^2/8 = \text{r.h.s. of (5.3) with } \kappa_2 = 1/8.$$

The upshot of this is that Theorem 5.1 implies that $\dim_H \text{Bad}_I \geq 1$. In turn, since Bad_I is a subset of $\mathbb{R}$, this implies that $\dim_H \text{Bad}_I = 1$.

The key feature exploited to check the conditions on the collection is the fact that rational numbers are well spaced. In higher dimensions, the appropriate analogue is the following lemma, the idea of which goes back to Davenport.

Lemma 5.3 *Let $n \geq 1$ be an integer and $k > 1$ be a real number. Let $E \subseteq \mathbb{R}^n$ be a convex set of n–dimensional Lebesgue measure*

$$|E| \leq \frac{1}{n!k^{-(n+1)}}.$$

Suppose that E contains $n + 1$ rational points $(p_i^{(1)}/q_i, \ldots, p_i^{(n)}/q_i)$ with $1 \leq q_i < k$, where $0 \leq i \leq n$. Then these rational points lie in some hyperplane.

Proof Suppose to the contrary that this is not the case. In that case, the rational points $(p_i^{(1)}/q_i, \ldots, p_i^{(n)}/q_i)$ where $0 \leq i \leq n$ are distinct. Consider the n–dimensional simplex Δ subtended by them, i.e. an interval when $n = 1$, a triangle when $n = 2$, a tetrahedron when $n = 3$ and so on. Clearly, Δ is a subset of E since E is convex. The volume $|\Delta|$ of the simplex times n factorial is equal to the absolute value of the determinant

$$\det = \begin{vmatrix} 1 & p_0^{(1)}/q_0 & \cdots & p_0^{(n)}/q_0 \\ 1 & p_1^{(1)}/q_1 & \cdots & p_1^{(n)}/q_1 \\ \vdots & \vdots & & \vdots \\ 1 & p_n^{(1)}/q_n & \cdots & p_n^{(n)}/q_n \end{vmatrix}.$$

As this determinant is not zero, it follows from the assumption made on the q_i that

$$n! \times |\Delta| = |\det| \geq \frac{1}{q_0 q_1 \cdots q_n} > k^{-(n+1)}.$$

Consequently, $|\Delta| > (n!)^{-1} k^{-(n+1)} \geq |E|$. This contradicts the fact that $\Delta \subseteq E$.

Of course, in one dimension this is exactly the spacing estimate used in the proof of Jarník's result above.

Lemma 5.3 serves to ensure that not too many rectangles are bad for the application of Theorem 5.1, but we need some way of ensuring that there are enough rectangles to begin with. Lemma 5.5 below accomplishes this and is proved using the following simple covering lemma.

Lemma 5.4 *Let (X, d) be the Cartesian product space of the metric spaces (X_1, d_1), ..., (X_t, d_t) and $\mathcal{F}$ be a finite collection of 'rectangles' $F = F(c; l_1, \ldots, l_t)$ with $c \in X$ and $l_1, \ldots, l_t$ fixed. Then there exists a disjoint sub-collection $\{F_m\}$ such that*

$$\bigcup_{F \in \mathcal{F}} F \subset \bigcup_m 3F_m.$$

Proof Let S denote the set of centres c of the rectangles in $\mathcal{F}$. Choose $c(1) \in S$ and for $k \geq 1$,

$$c(k+1) \in S \setminus \bigcup_{m=1}^{k} 2 F(c(m); l_1, \ldots, l_t)$$

as long as $S \setminus \bigcup_{m=1}^{k} 2 F(c(m); l_1, \ldots, l_t) \neq \emptyset$. Since #$S$ is finite, the process terminates and there exists $k_1 \leq$ #S such that

$$S \subset \bigcup_{m=1}^{k_1} 2 F(c(m); l_1, \ldots, l_t) .$$

By construction, any rectangle $F(c; l_1, \ldots, l_t)$ in the original collection $\mathcal{F}$ is contained in some rectangle $3 F(c(m); l_1, \ldots, l_t)$ and since $d_i(c_i(m), c_i(n)) > 2l_i$ for each $1 \leq i \leq t$ the chosen rectangles $F(c(m); l_1, \ldots, l_t)$ are clearly disjoint.

Lemma 5.5 *Let (X, d) be the Cartesian product of the metric spaces (X_1, d_1), ..., (X_t, d_t) and let (Ω, d, m) be a compact measure subspace of X. Let the measure m and the functions ρ_i satisfy conditions (B) to (D). Let k be sufficiently large. Then* *for any $\theta \in \mathbb{R}_+$ and for any rectangle F_n ($n \geq 1$) there exists a disjoint collection $\mathcal{C}(\theta F_n)$ of rectangles $2\theta F_{n+1}$ contained within θF_n satisfying (5.2) of Theorem 5.1.*

Proof Begin by choosing k large enough so that for any $i \in \{1, \ldots, t\}$,

$$\frac{\rho_i(k^n)}{\rho_i(k^{n+1})} \geq 4. \tag{5.7}$$

That this is possible follows from the fact that $\lambda_i^l(k) \to \infty$ as $k \to \infty$ (condition (B)). Take an arbitrary rectangle F_n and let $l_i(n) := \theta \rho_i(k^n)$. Thus $\theta F_n := F(c; l_1(n), \ldots, l_t(n))$. Consider the rectangle $T_n \subset \theta F_n$ where

$$T_n := F(c; l_1(n) - 2l_1(n+1), \ldots, l_t(n) - 2l_t(n+1)) .$$

Note that in view of (5.7) we have that $T_n \supset \frac{1}{2}\theta F_n$. Now, cover T_n by rectangles $2\theta F_{n+1}$ with centres in $\Omega \cap T_n$. By construction, these rectangles are contained in θF_n and in view of the Lemma 5.4 there exists a disjoint sub-collection $\mathcal{C}(\theta F_n)$ such that

$$T_n \subset \bigcup_{2\theta F_{n+1} \subset \mathcal{C}(\theta F_n)} 6\theta F_{n+1} .$$

Using the fact that rectangles of the same size centred at points of Ω have comparable m measure (condition (C)), it follows that

$$a\, m\left(\frac{1}{2}\theta F_n\right) \leq m(T_n) \leq \#\mathcal{C}(\theta F_n)\, b\, m(6\theta F_{n+1}) .$$

Using the fact that the measure m is doubling on rectangles (condition (D)), so that $m(\frac{1}{2}\theta F_n) \geq D^{-1}m(\theta F_n)$ and $m(6\theta F_{n+1}) \leq m(8\theta F_{n+1}) \leq D^3 m(\theta F_{n+1})$, it follows that

$$\#\mathcal{C}(\theta F_n) \geq \frac{a}{bD^4} \frac{m(\theta F_n)}{m(\theta F_{n+1})}.$$

With Theorem 5.1 and the lemmas, we can intersect the sets $\mathrm{Bad}(i_1, \ldots, i_n)$ with nice fractals. We begin with the case when $i_1 = \cdots = i_n = 1/n$.

Let Ω be a compact subset of $\mathbb{R}^n$ which supports a non-atomic, finite measure m. Let $\mathcal{L}$ denote a generic hyperplane of $\mathbb{R}^n$ and let $\mathcal{L}^{(\epsilon)}$ denote its ϵ-neighbourhood. We say that m is *absolutely* α-*decaying* if there exist strictly positive constants C, α, r_0 such that for any hyperplane $\mathcal{L}$, any $\epsilon > 0$, any $x \in \Omega$ and any $r < r_0$,

$$m\left(B(x, r) \cap \mathcal{L}^{(\epsilon)}\right) \leq C\left(\frac{\epsilon}{r}\right)^\alpha m(B(x, r)) .$$

This is a quantitative way of saying that the support of the measure does not concentrate on any hyperplane, and so a way of quantifying the statement that the set Ω

is sufficiently spread out in $\mathbb{R}^n$. If $\Omega \subset \mathbb{R}$ is supporting a measure m which satisfies (5.1), then it is relatively straightforward to show that m is absolutely δ–decaying. However, in higher dimensions one need not imply the other.

Inside the set Ω, we will define sets of weighted badly approximable numbers as follows. For $0 \le i_1, \ldots, i_n \le 1$ with $i_1 + \cdots + i_n = 1$, let

$$\mathrm{Bad}_\Omega(i_1, \ldots, i_n)$$

$$= \left\{ x \in \Omega : \max_{1 \le j \le n} \{ \|qx_i\|^{1/i_j} > c(x)q^{-1} \text{ for some } c(x) > 0, \text{ for all } q \in \kappa \right\}.$$

When $i_1 = \cdots = i_n = 1/n$, we will for brevity denote this set by $\mathrm{Bad}_\Omega(n)$.

Theorem 5.6 *Let Ω be a compact subset of $\mathbb{R}^n$ which supports a measure m satisfying condition (5.1) and which in addition is absolutely α–decaying for some $\alpha > 0$. Then*

$$\dim_H \mathrm{Bad}_\Omega(n) = \dim_H \Omega .$$

Proof The set $\mathrm{Bad}_\Omega(n)$ can be expressed in the form $\mathrm{Bad}^*(\mathcal{R}, \beta, \rho)$ with $\rho(r) = r^{-(1+\frac{1}{n})}$ and

$$X = (\mathbb{R}^n, d) , \quad J = \{((p_1, \ldots, p_n), q) \in \mathbb{N}^n \times \mathbb{N}\backslash\{0\}\} ,$$

$$\alpha = ((p_1, \ldots, p_n), q) \in J , \quad \beta_\alpha = q , \quad R_\alpha = (p_1/q, \ldots, p_n/q) .$$

Here d is standard sup metric on $\mathbb{R}^n$; $d(x, y) = \max\{d(x_1, y_1), \ldots, d(x_n, y_n)\}$. Thus balls $B(c, r)$ in $\mathbb{R}^n$ are genuinely cubes of sidelength $2r$.

We show that the conditions of Theorem 5.1 are satisfied. Clearly the function ρ satisfies condition (B) and we are given that the measure m supported on Ω satisfies condition (A) Conditions (C), (D) and (E) also follow from (5.1). Since the resonant sets R_α are all points, the condition $\dim_H(\cup_{\alpha \in J} R_\alpha) < \delta$ is satisfied by properties (iv) and (vii) of Hausdorff dimension. We need to establish the existence of the disjoint collection $\mathcal{C}(\theta B_n)$ of balls (cubes) $2\theta B_{n+1}$ where B_n is an arbitrary ball of radius $k^{-n(1+\frac{1}{n})}$ with centre in Ω. In view of Lemma 5.5, there exists a disjoint collection $\mathcal{C}(\theta B_n)$ such that

$$\#\mathcal{C}(\theta B_n) \ge \kappa_1 k^{(1+\frac{1}{n})\delta}; \tag{5.8}$$

i.e. (5.2) of Theorem 5.1 holds. We now verify that (5.3) is satisfied for any such collection.

We consider two cases.

Case 1: $n = 1$. The trivial spacing argument of (5.6) shows that any interval θB_n with $\theta := \frac{1}{2}k^{-2}$ contains at most one rational p/q with $k^n \le q < k^{n+1}$; i.e. $\alpha \in J(n + 1)$. Thus, for k sufficiently large

$$\text{l.h.s. of (5.3)} \leq 1 < \frac{1}{2} \times \text{r.h.s. of (5.8)} \,.$$

Hence (5.3) is trivially satisfied and Theorem 5.1 implies the desired result. As a special case, we have shown that the badly approximable numbers in the ternary Cantor set form a set of maximal dimension.

Case 2: $n \geq 2$. We will prove the theorem in the case that $n = 2$. There are no difficulties and no new ideas are required in extending the proof to higher dimensions. One just needs to apply Lemma 5.3 in higher dimensions.

Suppose that there are three or more rational points $(p_1/q, p_2/q)$ with $k^n \leq q < k^{n+1}$ lying within the ball/square θB_n. Now put $\theta = 2^{-1}(2k^3)^{-1/2}$. Then Lemma 5.3 implies that the rational points must lie on a line $\mathcal{L}$ passing through θB_n. Setting $\epsilon = 8\theta k^{-(n+1)\frac{3}{2}}$, it follows that

$$\text{l.h.s. of (5.3)} \leq \#\{2\theta B_{n+1} \subset C(\theta B_n) : 2\theta B_{n+1} \cap \mathcal{L} \neq \emptyset\}$$
$$\leq \#\{2\theta B_{n+1} \subset C(\theta B_n) : 2\theta B_{n+1} \subset \mathcal{L}^{(\epsilon)}\} \,.$$

Using that the balls $2\theta_{n+1}$ are disjoint and that the measure m is absolutely α-decaying, this is

$$\leq \frac{m(\theta B_n \cap \mathcal{L}^{(\epsilon)})}{m(2\theta B_{n+1})} \leq a^{-1}bC8^{\alpha}2^{-\delta}k^{\frac{2}{3}(\delta-\alpha)} \,.$$

On choosing k large enough, this becomes $\leq \frac{1}{2} \times$ r.h.s. of (5.8). Hence (5.3) is satisfied and Theorem 5.1 implies the desired result.

We now prove the result for general values of i_j, but under a more restrictive assumption on the underlying fractal.

Theorem 5.7 *For $1 \leq j \leq n$, let Ω_j be a compact subset of $\mathbb{R}$ which supports a measure m_j satisfying (5.1) with exponent δ_j. Let Ω denote the product set $\Omega_1 \times \cdots \times \Omega_n$. Then, for any n–tuple $(i_1, \ldots, i_n)$ with $i_j \geq 0$ and $\sum_{j=1}^{n} i_j = 1$,*

$$\dim_H \operatorname{Bad}_{\Omega}(i_1, \ldots, i_n) = \dim_H \Omega \,.$$

A simple application of the above theorem leads to following result.

Corollary 5.8 *Let K_1 and K_2 be regular Cantor subsets of $\mathbb{R}$. Then*

$$\dim_H ((K_1 \times K_2) \cap \operatorname{Bad}(i, j)) = \dim_H(K_1 \times K_2) = \dim_H K_1 + \dim_H K_2 \,.$$

Proof of Theorem 5.7. We shall restrict our attention to the case $n = 2$ and leave it for the reader to extend this to higher dimensions.

A relatively straightforward argument shows that $m := m_1 \times m_2$ is absolutely α–decaying on Ω with $\alpha := \min\{\delta_1, \delta_2\}$. In fact, more generally for $2 \leq j \leq n$, if each m_j is absolutely α_j–decaying on Ω_j, then $m := m_1 \times \ldots \times m_n$ is absolutely α–decaying on $\Omega = \Omega_1 \times \cdots \times \Omega_n$ with $\alpha = \min\{\alpha_1, \cdots, \alpha_n\}$.

Now let us write $\mathrm{Bad}(i, j)$ for $\mathrm{Bad}(i_1, i_2)$ and without loss of generality assume that $i < j$. The case $i = j$ is already covered by Theorem 4 since m is absolutely α–decaying on Ω and clearly satisfies (5.1). The set $\mathrm{Bad}_\Omega(i, j)$ can be expressed in the form $\mathrm{Bad}^*(\mathcal{R}, \beta, \rho_1, \rho_2)$ with $\rho_1(r) = r^{-(1+i)}$, $\rho_2(r) = r^{-(1+j)}$ and

$$X = \mathbb{R}^2, \quad \Omega = \Omega_1 \times \Omega_2, \quad J = \{((p_1, p_2), q) \in \mathbb{N}^2 \times \mathbb{N}\backslash\{0\}\},$$

$$\alpha = ((p_1, p_2), q) \in J, \quad \beta_\alpha = q, \quad R_\alpha = (p_1/q, p_2/q).$$

The functions ρ_1, ρ_2 satisfy condition (B) and the measures m_1, m_2 satisfy (5.1). Also note that $\dim_H(\cup_{\alpha \in J} R_\alpha) = 0$ since the union in question is countable. We need to establish the existence of the collection $C(\theta F_n)$, where each F_n is an arbitrary closed rectangle of size $2k^{-n(1+i)} \times 2k^{-n(1+j)}$ with centre c in Ω. By Lemma 5.5, there exists a disjoint collection $C(\theta F_n)$ of rectangles $2\theta F_{n+1} \subset \theta F_n$ such that

$$\#C(\theta F_n) \geq \kappa_1 k^{(1+i)\delta_1} k^{(1+j)\delta_2}; \tag{5.9}$$

i.e. (5.4) of Theorem 5.2 is satisfied. We now verify that (5.5) is satisfied for any such collection. With $\theta = 2^{-1}(2k^3)^{-1/2}$, the Lemma 5.3 implies that

$$\text{l.h.s. of } (5.5) \leq \#\{2\theta F_{n+1} \subset C(\theta F_n) : 2\theta F_{n+1} \cap \mathcal{L} \neq \emptyset\}, \tag{5.10}$$

where $\mathcal{L}$ is a line passing through θF_n. Consider the thickening $T(\mathcal{L})$ of $\mathcal{L}$ obtained by placing rectangles $4\theta F_{n+1}$ centred at points of $\mathcal{L}$; that is, by 'sliding' a rectangle $4\theta F_{n+1}$, centred at a point of $\mathcal{L}$, along $\mathcal{L}$. Then, since the rectangles $2\theta F_{n+1} \subset C(\theta F_n)$ are disjoint,

$$\#\{2\theta F_{n+1} \subset C(\theta F_n) : 2\theta F_{n+1} \cap \mathcal{L} \neq \emptyset\} \tag{5.11}$$
$$\leq \#\{2\theta F_{n+1} \subset C(\theta F_n) : 2\theta F_{n+1} \subset T(\mathcal{L})\}$$
$$\leq \frac{m(T(L) \cap \theta F_n)}{m(2\theta F_{n+1})}.$$

Without loss of generality we can assume that $\mathcal{L}$ passes through the centre of θF_n. To see this, suppose that $m(T(\mathcal{L}) \cap \theta F_n) \neq 0$ since otherwise there is nothing to prove. Then, there exists a point $x \in T(\mathcal{L}) \cap \theta F_n \cap \Omega$ such that

$$T(\mathcal{L}) \cap \theta F_n \subset 2\theta F_n' \cap T'(\mathcal{L}').$$

Here F_n' is the rectangle of size $k^{-n(1+i)} \times k^{-n(1+j)}$ centred at x, $\mathcal{L}'$ is the line parallel to $\mathcal{L}$ passing through x and $T'(\mathcal{L}')$ is the thickening obtained by 'sliding' a rectangle $8\theta F_{n+1}$ centred at x, along $\mathcal{L}'$. Then the following argument works just as well on $2\theta F_n' \cap T'(\mathcal{L}')$.

Let Δ denote the slope of the line $\mathcal{L}$ and assume that $\Delta \geq 0$. The case $\Delta < 0$ can be dealt with similarly. By moving the rectangle θF_n to the origin, straightforward geometric considerations lead to the following facts:

(F1)

$$T(\mathcal{L}) = \mathcal{L}^{(\epsilon)} \text{ where } \epsilon = \frac{4\theta \left(k^{-(n+1)(1+j)} + \Delta k^{-(n+1)(1+i)}\right)}{\sqrt{1 + \Delta^2}},$$

(F2) $T(\mathcal{L}) \cap \theta F_n \subset F(c; l_1, l_2)$ where $F(c; l_1, l_2)$ is the rectangle with the same centre c as F_n and of size $2l_1 \times 2l_2$ with

$$l_1 = \frac{\theta}{\Delta} \left(k^{-n(1+j)} + 4k^{-(n+1)(1+j)} + \Delta k^{-(n+1)(1+i)}\right) \text{ and } l_2 = \theta k^{-n(1+j)}.$$

The asymmetrical shape of the sliding rectangle adds tremendously to the technical calculations from now on. However, we can in fact estimate the right hand side of (5.11) by considering two cases, depending on the magnitude of Δ. Throughout, let a_i, b_i denote the constants associated with the measure m_i and condition (5.1) and let

$$\varpi = 3 \left(\frac{4b_1 b_2}{\kappa_1 a_1 a_2 2^{\delta_1 + \delta_2}}\right)^{1/\delta_1}.$$

Case 1: $\Delta \geq \varpi k^{-n(1+j)}/k^{-n(1+i)}$. In view of (F2) above, we trivially have that

$$m(\theta F_n \cap T(\mathcal{L})) \leq m(F(c; l_1, l_2)) \leq b_1 b_2 l_1^{\delta_1} l_2^{\delta_2}.$$

It follows that

$$
\begin{aligned}
\frac{m(T(L) \cap \theta F_n)}{m(2\theta F_{n+1})} &\leq \frac{b_1 b_2 l_1^{\delta_1} l_2^{\delta_2}}{a_1 a_2 (2\theta)^{\delta_1 + \delta_2} k^{-(n+1)(1+j)\delta_1} k^{-(n+1)(1+i)\delta_2}} \\
&\leq \frac{b_1 b_2}{a_1 a_2 2^{\delta_1 + \delta_2}} \left(\frac{1}{\varpi} + \frac{1}{\varpi k^{1+j}} + \frac{1}{k^{1+i}}\right)^{\delta_1} k^{(1+j)\delta_1 + (1+i)\delta_2} \\
&\leq \frac{b_1 b_2}{a_1 a_2 2^{\delta_1 + \delta_2}} \left(\frac{3}{\varpi}\right)^{\delta_1} k^{(1+j)\delta_1 + (1+i)\delta_2} = \frac{\kappa_1}{4} k^{(1+j)\delta_1 + (1+i)\delta_2}.
\end{aligned}
$$

Case 2: $0 \leq \Delta < \varpi k^{-n(1+j)}/k^{-n(1+i)}$. By Lemma 5.4, there exists a collection $\mathcal{B}_n$ of disjoint balls B_n with centres in $\theta F_n \cap \Omega$ and radii $\theta k^{-n(1+j)}$ such that

$$\theta F_n \cap \Omega \subset \bigcup_{B_n \in \mathcal{B}_n} 3B_n.$$

Since $i < j$, it is easily verified that the disjoint collection $\mathcal{B}_n$ is contained in $2\theta F_n$ and thus $\#\mathcal{B}_n \leq m(2\theta F_n)/m(B_n)$. It follows that

$$m(\theta F_n \cap T(\mathcal{L})) \leq m\left(\cup_{B_n \in \mathcal{B}_n} 3B_n \cap T(\mathcal{L})\right) \leq \#\mathcal{B}_n\, m(3B_n \cap T(\mathcal{L})).$$

Applying (F1) and subsequently the fact that m is absolutely α-decaying, this is

$$\leq \frac{m(2\theta F_n)}{m(B_n)} m\left(3B_n \cap \mathcal{L}^{(\epsilon)}\right) \leq m(2\theta F_n) \frac{m(3B_n)}{m(B_n)} \left(\frac{\epsilon}{3\theta k^{-n(i+j)}}\right)^\alpha$$

Now notice that

$$\frac{\epsilon}{3\theta k^{-n(i+j)}} \leq \frac{4}{3}(k^{-(1+j)} + \varpi k^{-(1+i)}).$$

Hence, for k sufficiently large we have

$$\frac{m(T(\mathcal{L}) \cap \theta F_n)}{m(2\theta F_{n+1})} \leq \frac{\kappa_1}{4} k^{(1+j)\delta_1} k^{(1+i)\delta_2}.$$

On combining the above two cases, we have

$$\text{l.h.s. of (5.5)} \leq \frac{m(T(\mathcal{L}) \cap \theta F_n)}{m(2\theta F_{n+1})} \leq \frac{\kappa_1}{4} k^{(1+j)\delta_1} k^{(1+i)\delta_2} = \frac{1}{4} \times \text{l.h.s. of (5.9).}$$

Hence (5.5) is satisfied and Theorem 5.2 implies the desired result.

We give a few remarks on the results above. Firstly, an alternative approach using homogeneous dynamics is also known due to Kleinbock and Weiss [33]. This approach is less versatile, as it only works in real, Euclidean space, but the conditions on the measure are slightly less restrictive, so in this respect their result is stronger.

Secondly, it would be really nice if the approach could give a result on numbers badly approximable by algebraic numbers. Unfortunately, the known spacing estimates for algebraic numbers are not good enough to get the naive approach to work (more on this in the next section). One could hope that proving a result on badly approximable vectors on the Veronese curves would work, but unfortunately Lemma 5.3 does not allow us to control the direction of the hyperplane containing the rational points. If this hyperplane is close to tangential to the curve, the approach used above will not work, so more input is needed. In fact, the problem has now been resolved for the full Veronese curve in 2 dimensions by Badziahin and Velani [2] and in higher dimensions by Beresnevich [5]. To the knowledge of the author, the problem of intersecting the sets with fractals remain unsolved.

Thirdly, our failure in proving Schmidt's conjecture with this approach was due to the fact that we could not construct a measure on $\text{Bad}(i, j)$ satisfying the conditions of Theorem 5.1. Since the publication of the paper, the conjecture has been settled, and in fact An [1] proved that the sets $\text{Bad}(i, j)$ are winning for the so-called Schmidt game [45]. This implies that they are stable under countable intersection. We proceed with a discussion of Schmidt games in one dimension and leave it as an exercise to extend this to higher dimensions.

Definition 5.9 Let $F \subseteq \mathbb{R}$, and let $\alpha, \beta \in (0, 1)$. The *Schmidt game* is played by two players, Black and White, according to the following rules:

1. Black picks a closed interval B_1 of length r.
2. White picks a closed interval $W_1 \subseteq B_1$ of length αr.
3. Black picks a closed interval $B_2 \subseteq W_1$ of length $\beta \alpha r$.
4. And so on…

By Banach's fixpoint theorem, $\cap_i B_i$ consists of a single point, x say. If $x \in F$, White wins the game. Otherwise, Black wins.

By requiring that the initial ball is chosen in a particular way, we can use the above for bounded sets without breaking the game.

We are concerned with winning strategies for the game. In particular, we will prove that if White has a winning strategy, then the set F is large.

Definition 5.10 A set F is said to be (α, β)-winning if White can always win the Schmidt game with these parameters. A set F is said to be α-winning if it is (α, β)-winning for any $\beta \in (0, 1)$.

We will prove the following theorems.

Theorem 5.11 *An α-winning set F has Hausdorff dimension* 1.

Theorem 5.12 *If (F_i) is a sequence of α-winning sets, then $\cap_i F_i$ is also α-winning.*

Proof of Theorem 5.11. We suppose without loss of generality that $r = 1$. For ease of computation, we will also assume that $\beta = 1/N$ for an integer $N > 1$. Given an interval W_k in the game, we may partition this set into N (essentially) disjoint intervals. We will restrict the possible choices that Black can make by requiring that she picks one of these. By requiring that β was slightly smaller than $1/N$, we could ensure that the intervals were properly disjoint. We will continue our calculations with this assumption, even though we should strictly speaking add a little more technicality to the setup.

Given that White plays according to a winning strategy, we find disjoint paths through the game depending on the choices made by Black. In other words, for each possible resulting element in F, we find a unique sequence with elements in $\{0, \ldots, N - 1\}$ and *vice versa*, for each such sequence, we obtain a resulting element. Hence, this particular subset of F may be mapped onto the unit interval by thinking of the sequence from the game as a sequence of digits in the base N-expansion of a number between 0 and 1. To sum up, we have constructed a surjective function

$$g : F^* \to [0, 1],$$

where $F^* \subseteq F$. We extend this to a function on arbitrary subsets of $\mathbb{R}$ by setting $g(A) = g(A \cap F^*)$.

Now, let $\{U_i\}$ be a cover of F^* with U_i having diameter ρ_i. Then, with $\mathcal{L}$ denoting the outer Lebesgue measure,

$$\sum_{i=1}^{\infty} \mathcal{L}(g(U_i)) \geq \mathcal{L}\left(\bigcup_{i=1}^{\infty} g(U_i)\right) \geq \mathcal{L}([0, 1]) = 1.$$

Let $\omega > 0$ be so small that any interval of length $\omega(\alpha\beta)^k$ intersects at most two of the generation k intervals chosen by Black, i.e. any of the intervals $B_k(j_1, \ldots, j_k)$ where $j_i \in \{0, \ldots, N-1\}$. Even in higher dimensions, $\omega = 2/\sqrt{3} - 1$ will do nicely. Finally, define integers

$$k_i = \left[\frac{\log(2\omega^{-1}\rho_i)}{\log \alpha\beta}\right].$$

If ρ_i is sufficiently small, then $k_i > 0$ and $\rho_i < \omega(\alpha\beta)^{k_i}$. Hence, the interval U_i intersects at most two of the generation k_i-intervals chosen by Black. The image of such an interval under g is evidently an interval of length N^{-k_i}, so since there are no more than two of them, $\mathcal{L}(g(U_i)) \leq 2N^{-k_i}$. Summing up over i, we find that

$$1 \leq \sum_{i=1}^{\infty} \mathcal{L}(g(U_i)) \leq \sum_{i=1}^{\infty} 2N^{-k_i} \leq K \sum_{i=1}^{\infty} \rho_i^{\frac{\log N}{|\log(\alpha\beta)|}},$$

where $K > 0$ is explicitly computable in terms of N, α, β and ω. Nonetheless, we have obtained a positive lower bound on the $\frac{\log N}{|\log(\alpha\beta)|}$-length of an arbitrary cover of F^* with small enough sets. It follows that

$$\dim_{\mathrm{H}}(F) \geq \frac{\log N}{|\log(\alpha\beta)|} = \frac{|\log \beta|}{|\log \alpha| + |\log \beta|}.$$

The result now follows on letting $\beta \to 0$.

Proof of Theorem 5.12. White plays according to different strategies at different stages of the game. Explicitly, for α and β fixed, in the first, third, fifth etc. move, White plays according to a $(\alpha, \alpha\beta\alpha; E_1)$-winning strategy, i.e. a strategy for the $(\alpha, \alpha\beta\alpha; E_1)$ for which White is guaranteed to win the game. Since $\rho(B_{l+1}) = \alpha\beta\alpha\rho(B_{l-1})$, this is a valid strategy, and hence the resulting $x \in E_1$. Along the second, sixth, tenth etc. move, White plays according to a $(\alpha, \alpha(\beta\alpha)^3; E_2)$-winning strategy. This is equally valid, and ensures that $x \in E_2$.

In general, in the k'th move with $k \equiv 2^{l-1} \pmod{2)^l}$, White moves as if he was playing the $(\alpha, \alpha(\beta\alpha)^{2^l-1}; E_l)$-game. This ensures that the resulting element x is an element of E_l for any l.

A positional strategy is a strategy which may be chosen by looking only at the present state of the game without taking previous moves into account. In An's proof, the strategy chosen by White is not positional (at least it appears not to be). This is a little annoying, as it was shown by Schmidt that any winning set admits a positional strategy. Of course, this proof depends on the well-ordering principle, and so ultimately on the axiom of choice. Describing a positional winning strategy may hence not be that easy.

6   Well Approximable Elements

In this section, we return to where we started, namely to Khintchine's theorem and to fractals arising from continued fractions. We will address three problems. The first is the problem of the size of the exceptional sets in Khintchine's theorem. This will be resolved using a technique due to Beresnevich and Velani known as the mass transference principle. The second is concerned with the ternary Cantor set and the Diophantine properties of elements in it. We will discuss the possibility of getting a Khintchine type theorem for this set and also give a quick-and-dirty argument, stating that most numbers in the set are not ridiculously well approximable by algebraic numbers. Finally, we will remark on some fractal properties which can be used in the study of Littlewood's conjecture.

Initially, we begin with a discussion of the exceptional sets arising from Khintchine's theorem. The Hausdorff dimension of the null sets in the case of convergence was originally calculated by Jarník [30] and independently by Besicovitch [10]. Various new methods were introduced during the last century, with the notion of ubiquitous systems being a key concept in recent years. Ubiquity was introduced (or at least named) by Dodson, Rynne and Vickers [20] and put in a very general form by Beresnevich, Dickinson and Velani [8]. A complete discussion of ubiquity will not be given here, but the reader is strongly encouraged to look up the paper Beresnevich, Dickinson and Velani.

Even more recently (this century), it was observed by Beresnevich and Velani [7] that under relatively mild assumptions on a *limsup* set, one may transfer a zero–one law for such a set to a zero–infinity law for Hausdorff measures, at least in the case of full measure. In the cases considered in these notes, the converse case of measure zero is easy. Note that the measure zero case is not always the easiest! One can cook up problems where the convergence case of a Khintchine type theorem is the difficult part. The sets considered in these notes however all fall within the category where divergence is the difficult problem.

At the heart of the observation of Beresnevich and Velani is the following theorem, usually called the mass transference principle. To state it, we will need a little notation. For a ball $B = B(x, r) \subseteq \mathbb{R}^n$ and a dimension function f, we define

$$B^f = B(x, f(r)^{1/n}),$$

the ball with the same centre but with its radius adjusted according to the dimension function and the dimension of the ambient space. As usual, for $f(r) = r^k$, we denote B^f by B^k.

Theorem 6.1 *Let $\{B_i\}$ be a sequence of balls in $\mathbb{R}^n$ with $r(B_i) \to 0$ as $i \to \infty$. Let f be a dimension function such that $r^{-n} f(r)$ is monotonic. Suppose that*

$$\mathcal{H}^n(B \cap \limsup B_i^f) = \mathcal{H}^n(B),$$

for any ball $B \subseteq \mathbb{R}^n$. Then, for any ball $B \subseteq \mathbb{R}^n$,

$$\mathcal{H}^f(B \cap \limsup B_i^n) = \mathcal{H}^f(B),$$

Notice that the first requirement just says that the set $\limsup B_i^f$ is full with respect to Lebesgue measure, as $\mathcal{H}^n$ is comparable with the Lebesgue measure. Note also, that if $r^{-n} f(r) \to \infty$ as $r \to 0$, $\mathcal{H}^f(B) = \infty$, so a re-statement of the conclusion of the theorem would be as follows: Suppose that a *limsup* set of balls is full with respect to Lebesgue measure. Then a *limsup* set of appropriately scaled balls is of infinite Hausdorff measure.

The mass transference principle is valid in a more general setting of certain metric spaces. As was the case in the framework of badly approximable sets, the metric space must support a natural measure, which in this case should be a Hausdorff measure. This is the case for the Cantor set with the Hausdorff log 2/log 3-measure, and the mass transference principle is exactly the same if one reads log 2/log 3 for n everywhere.

Finally, the reader will note that the mass transference principle in the present form only works for *limsup* sets of balls. If one were to consider linear forms approximation as we did in Sect. 4, the *limsup* set would be built from tubular neighbourhoods of hyperplanes, and for the more general setting of systems of linear forms from tubular neighbourhoods of lower dimensional affine subspaces. This can be overcome by a slicing technique, also developed by Beresnevich and Velani [6].

We will not go into details on the higher dimensional variant here, nor will we prove the mass transference principle. Instead, we will deduce the original Jarník–Besicovitch theorem from Khintchine's theorem.

Theorem 6.2 *Let f be a dimension function and let $\psi : \mathbb{N} \to \mathbb{R}_{\geq 0}$ be some function with $q^2 f(\psi(q))$ decreasing. Then,*

$$\mathcal{H}^f \left\{ x \in [0,1] : \left| x - \frac{p}{q} \right| < \psi(q) \text{ for infinitely many } (p, q) \in \mathbb{Z} \times \mathbb{N} \right\},$$

is zero or infinity according to whether the series $\sum q f(\psi(q))$ converges or diverges.

Proof The convergence half is the usual covering argument, which we omit. For the divergence part, we apply the mass transference principle. The balls are indexed by rational numbers, with $B_{p/q} = B(p/q, \psi(q))$ and $k = 1$, so the *limsup* set of the conclusion of Theorem 6.1 is just the set from the statement of the theorem. Hence, it suffices to prove that $\limsup B_{p/q}^f$ is full with respect to Lebesgue measure, provided the series in question diverges. But this is just the statement of the original Khintchine's theorem.

An easy corollary of this statement tells us, that for $\psi(q) = q^{-v}$, the upper bound of $2/v$ obtained on the Hausdorff dimension of the above set in Sect. 3 is sharp. In fact, in Khintchine's theorem, the requirement that $q^2 \psi(q)$ is monotonic can be relaxed substantially to the requirement that $\psi(q)$ is monotonic, which in turn gives us the Hausdorff measure at the critical dimension (it is infinite) by the above argument. The latter result could be deduced directly from Dirichlet's theorem, as the set of all numbers evidently is full, but only for the special case of the approximating function $\psi(q) = q^{-v}$.

We leave it for the reader to explore applications of the mass transference principle (there are many). The point we want to make is that once a zero–one law for a natural measure is known (Lebesgue in the case of the real numbers), it is usually a straightforward matter of applying the mass transference principle to get a Hausdorff measure variant of the known result. In other words, it is natural to look for a zero–one law for the natural measure and deduce the remainder of the metrical theory from this result.

We now consider the ternary Cantor set. Recall that the natural measure μ constructed in Sect. 3 on this set has the nice decay property, that for $\delta = \log 2/\log 3$ and $c_1, c_2 > 0$,

$$c_1 r^{\delta} \leq \mu([c - r, c + r]) \leq c_2 r^{\delta} \tag{6.1}$$

for all $c \in \mathcal{C}$ and $r > 0$ small enough. We used this property in the preceeding section as well. It is easy to see that any non-atomic measure supported on $\mathcal{C}$ satisfying hypothesis (6.1) must also satisfy

$$\mu([c - \epsilon r, c + \epsilon r]) \leq c_3 \epsilon^{\delta} \mu([c - r, c + r]), \tag{6.2}$$

for some $c_3 > 0$, whenever r and ϵ are small and $c \in \mathbb{R}$. The inequality in (6.2) is the statement that the measure is absolutely δ-decaying, which was also used in the preceeding section. In fact, this measure can also be seen to be the restriction of the Hausdorff δ-measure to $\mathcal{C}$, so we are within the framework where the mass transference principle can be applied.

Levesley, Salp and Velani [37] proved a zero–one law (and deduced the corresponding statement for Hausdorff measures) for the set

$$W_{\mathcal{C}} = \left\{ x \in \mathcal{C} : \left| x - \frac{p}{3^n} \right| < \psi(3^n) \text{ for infinitely many } (p, n) \in \mathbb{Z} \times \mathbb{N} \right\}.$$

Note the restriction on the approximating rationals. They are all rationals whose denominator is a power of 3, and so are the endpoints in the usual construction of the set. This is of course not satisfactory, as there are other rationals in the Cantor set, e.g. $1/4$. Nevertheless, we do not at present know a full zero–one law for approximation of elements in the Cantor set by rationals in the Cantor set. Their result is the following.

Theorem 6.3

$$\mu(W_{\mathcal{C}}) = \begin{cases} 0, & \sum_{n=1}^{\infty} (3^n \psi(3^n))^{\delta} < \infty, \\ 1, & \sum_{n=1}^{\infty} (3^n \psi(3^n))^{\delta} = \infty. \end{cases}$$

We make a few comments on the proof, as it is a good model for many proofs of zero–one laws for *limsup* sets. The convergence part is usually proved by a covering argument as we have seen. For the divergence part, we give an outline of the method used by Levesley, Salp and Velani. From Sect. 4, recall the converse to the Borel–

Cantelli lemma given in Lemma 4.11. By this lemma, it would suffice to prove that the sets forming the *limsup* set are pairwise independent to ensure full measure. However in general, we have no reason to suspect that these sets are pairwise independent, which is always a problem. However, if they satisfy the weaker condition of quasi-pairwise independence (see below), the first part of the lemma will give positive measure. This can subsequently be inflated in a number of ways. One possibility is to look for an underlying invariance for an ergodic transformation. Another is to apply the following local density condition.

Lemma 6.4 *Let μ be a finite, doubling Borel measure supported on a compact set $X \subseteq R^k$, and let $E \subseteq X$ be a Borel set. Suppose that there are constants $r_0, c > 0$, such that for any ball $B = B(x, r)$ with $x \in X$ and $r < r_0$,*

$$\mu(E \cap B) \geq c\mu(B).$$

Then E is full in X with respect to μ.

This result is a consequence of the Lebesgue density theorem.

In order to prove a zero–one law, one now attempts to verify the conditions of Lemma 6.4, with E being the *limsup* set, using Lemma 4.11. In other words, for the *limsup* set

$$\Lambda = \lim \sup E_n,$$

there is a constant $c > 0$, such that for any sufficiently small ball B centred in X,

$$\frac{\mu((E_m \cap B) \cap (E_n \cap B))}{\mu(B)} \leq c \, \frac{\mu(E_m \cap B)}{\mu(B)} \, \frac{\mu(E_n \cap B)}{\mu(B)} \tag{6.3}$$

whenever $m \neq n$. The probability measure used in Lemma 4.11 is the normalised restriction of μ to B. Just inserting the above estimate proves that $\mu(\Lambda \cap B) \geq c^{-1}\mu(B)$, whence Λ is full within X. In other words, it suffices to prove local pairwise quasi-independence of events in the above sense.

For the result of Levesley, Salp and Velani, one considers the subset of $W_{\mathcal{C}}(\psi)$ which is the *limsup* set of the sets

$$E_n = \bigcup_{\substack{0 \leq p \leq 3^n \\ 3 \nmid p}} B\left(\frac{p}{3^n}, \psi(3^n)\right) \cap \mathcal{C}.$$

After making some preliminary reductions, it is then possible to prove (6.3) by splitting up into the cases when m and n are pretty close (in which case intersection on the left hand side is empty) and the case when they are pretty far apart, where some clever counting arguments and the specific form of the measure is needed. The point we want to make is not in the details, but rather in the methods applied.

Of course, applying the mass transference principle immediately gives a condition for the Hausdorff measure of the set $W_{\mathcal{C}}$ to be infinite. Combining this with a covering argument for the convergence case, Levesley, Salp and Velani obtained the following theorem in full.

Theorem 6.5 *Let f be a dimension function with $r^{-\delta}f(r)$ monotonic. Then,*

$$\mathcal{H}^f(W_{\mathcal{C}}) = \begin{cases} 0, & \sum_{n=1}^{\infty} 3^{\delta n} f(\psi(3^n)) < \infty. \\ \mathcal{H}^f(\mathcal{C}), & \sum_{n=1}^{\infty} 3^{\delta n} f(\psi(3^n)) = \infty. \end{cases}$$

We now consider the approximation of elements of $\mathcal{C}$ by elements of $\mathbb{A}_n$, where the quality of approximation is measured in terms of the height of the approximating number. The present argument is from [35]. Of course, we cannot hope to get a Khintchine type result by the methods above, as we do not expect there to be any algebraic irrational elements in $\mathcal{C}$. In fact, we can say very little, and we are only able to get a convergence result. We proceed to give a quick argument, which is not best possible, but relatively short.

Let $\psi : \mathbb{R}_{\geq 1} \to \mathbb{R}_+$. We define the set

$$\mathcal{K}_n^*(\psi; \mathcal{C}) = \{x \in \mathcal{C} : |x - \alpha| < \psi(H(\alpha)) \text{ for infinitely many } \alpha \in \mathbb{A}_n\}. \quad (6.4)$$

Theorem 6.6 *Let $\mathcal{C}$ be the ternary Cantor set and let $\delta = \log 2/\log 3$. Suppose that $\psi : \mathbb{R}_{\geq 1} \to \mathbb{R}_{\geq 0}$ satisfies either*

$$\sum_{r=1}^{\infty} r^{2n\delta-1}\psi(r)^{\delta} < \infty \text{ and } \psi \text{ is non-increasing} \quad or \quad \sum_{r=1}^{\infty} r^n \psi(r)^{\delta} < \infty.$$

Then

$$\mu\left(\mathcal{K}_n^*(\psi; \mathcal{C})\right) = 0.$$

The result is almost surely not sharp. We first prove that the convergence of the first series ensures that the measure is zero. This is by far the most difficult part of the proof. We will use a bound on the distance between algebraic numbers, which is in a sense best possible. If α and β are distinct real algebraic numbers of degree at most n, then

$$|\alpha - \beta| \geq c_4 H(\alpha)^{-n} H(\beta)^{-n}, \quad (6.5)$$

where the constant $c_4 > 0$ depends solely on n. The result can be found in Bugeaud's book [13] as a special case of corollary A.2, where the explicit form of the constant c_4 is also given. It generalises (5.6), which is the same estimate for rational numbers between 0 and 1. In the case of rational numbers, the spacing distribution is much more well-behaved than for real algebraic numbers of higher degree, and for this reason, Theorem 6.6 is almost certainly not as sharp as it could be. Nonetheless, as remarked in [13], the estimate in (5.6) is in some sense best possible.

If for some $k \in \mathbb{N}$, $2^k \le H(\alpha), H(\beta) < 2^{k+1}$, (6.5) implies that $|\alpha - \beta| > \frac{1}{2}c_4 2^{-2n(k+1)}$. Consequently, for distinct real algebraic numbers α_i with $2^k \le H(\alpha_i) < 2^{k+1}$, the intervals $[\alpha_i - \frac{1}{4}c_4 2^{-2n(k+1)}, \alpha_i + \frac{1}{4}c_4 2^{-2n(k+1)}]$ are disjoint.

Let $k \in \mathbb{N}$. We will show that as $k \to \infty$,

$$\max_{2^k \le r < 2^{k+1}} \frac{\psi(r)}{4^{-1}c_4 2^{-2n(k+1)}} = o(1). \tag{6.6}$$

In other words, the ratio tends to 0 as k tends to infinity. Indeed, suppose to the contrary that there is a $c_5 > 0$ and a strictly increasing sequence $\{k_i\}_{i=1}^{\infty} \subseteq \mathbb{N}$ such that for any $i \in \mathbb{N}$

$$\max_{2^{k_i} \le r < 2^{k_i+1}} \frac{\psi(r)}{4^{-1}c_4 2^{-2n(k_i+1)}} > c_5.$$

By the convergence assumption of the theorem together with Cauchy's condensation criterion and the monotonicity of ψ,

$$\sum_{k=1}^{\infty} 2^{2n(k+1)\delta} \psi(2^k)^{\delta} = 2^{2n\delta} \sum_{k=1}^{\infty} \left(2^{2kn}\psi(2^k)\right)^{\delta} < \infty.$$

On the other hand, as ψ is non-increasing,

$$\sum_{k=1}^{\infty} 2^{2n(k+1)\delta}\psi(2^k)^{\delta} \ge 4^{-\delta}c_4^{\delta} \sum_{i=1}^{\infty} \left(\max_{2^{k_i} \le r < 2^{k_i+1}} \frac{\psi(r)}{4^{-1}c_4 2^{-2n(k_i+1)}}\right)^{\delta}$$

$$\ge 4^{-\delta}c_4^{\delta}c_5^{\delta} \sum_{i=1}^{\infty} 1 = \infty,$$

which is the desired contradiction.

Consider the sets

$$E_k = \bigcup_{\substack{\alpha \in \mathbb{A}_n \\ 2^k \le H(\alpha) < 2^{k+1}}} [\alpha - \psi(H(\alpha)), \alpha + \psi(H(\alpha))].$$

Clearly, for k large enough

$$\mu(E_k) \le \sum_{\substack{\alpha \in \mathbb{A}_n \\ 2^k \le H(\alpha) < 2^{k+1}}} \mu([\alpha - \psi(H(\alpha)), \alpha + \psi(H(\alpha))])$$

$$\le c_3 c_4^{\delta} 4^{-\delta} 2^{2n(k+1)\delta} \psi(2^k)^{\delta}$$

$$\sum_{\substack{\alpha \in \mathbb{A}_n \\ 2^k \le H(\alpha) < 2^{k+1}}} \mu([\alpha - \frac{1}{4}c_4 2^{-2n(k+1)}, \alpha + \frac{1}{4}c_4 2^{-2n(k+1)}]),$$

where we have used (6.2) and (6.6). The intervals in the final sum are disjoint. Hence, the sum of their measure is bounded from above by the measure of K, which is equal to 1. We have shown that for $k \geq k_0$,

$$\mu(E_k) \leq c_3 c_4^\delta 4^{-\delta} 2^{2n(k+1)\delta} \psi(2^k)^\delta.$$

To complete the proof of this case, we note that $\mathcal{K}_n^*(\psi; K)$ is the set of points falling in infinitely many of the E_k. But

$$\sum_{k=k_0}^\infty \mu(E_k) \leq c_3 c_4^\delta 4^{-\delta} \sum_{k=k_0}^\infty 2^{2n(k+1)\delta} \psi(2^k)^\delta = c_3 c_4^\delta 4^{-\delta} 2^{2n\delta} \sum_{k=k_0}^\infty 2^{2nk\delta} \psi(2^k)^\delta.$$

Using Cauchy's condensation criterion and the convergence assumption of the theorem, the latter series converges. Hence, the Borel–Cantelli lemma implies the theorem.

To show that the convergence of the second series is sufficient to ensure zero measure, we note that

$$\#\{\alpha \in \mathbb{A}_n : \alpha \in [0, 1], H(\alpha) = H\} \leq n(n+1)(2H+1)^n. \tag{6.7}$$

By (6.1), for any such α, we have $\mu([\alpha - \psi(H); \alpha + \psi(H)]) \leq c_6 \psi(H)^\delta$ for some $c_6 > 0$. Elements of $\mathcal{K}_n^*(\psi; K)$ fall in infinitely many of these intervals, and as

$$\sum_{H=1}^\infty \sum_{\substack{\alpha \in \mathbb{A}_n \\ \alpha \in [0,1] \\ H(\alpha) = H}} \mu([\alpha - \psi(H); \alpha + \psi(H)]) \leq n(n+1)c_6 \sum_{H=1}^\infty (2H+1)^n \psi(H)^\delta,$$

which converges by assumption, the measure of $\mathcal{K}_n^*(\psi; K)$ is zero by the Borel–Cantelli lemma. $\square$

It is possible to prove a stronger result using homogeneous dynamics. This was done by Kleinbock, Lindenstrauss and Weiss [34], but the present result has the advantage of being relatively simple to prove.

The final thing, which we will touch upon in these notes, is a result on the Littlewood conjecture, which uses the Fourier dimension, which we defined in Sect. 3, but did not use for anything. We will sketch a proof of the following result, which is a partial result of [28].

Theorem 6.7 *Let $\{\alpha_i\} \subseteq$ Bad be a countable set of badly approximable numbers. The set of $\beta \in$ Bad for which all pairs (α_i, β) satisfy the Littlewood conjecture is of Hausdorff dimension 1.*

For a single α_i, this result was also proven by Pollington and Velani [42] by a similar, but slightly more complicated method. It could also be deduced from homogeneous dynamics, but the present method is different, and in fact gives a stronger result. However, it falls short of anything near the seminal result of Einsiedler, Katok and Lindenstrauss [23].

Note that unless both α and β are badly approximable, the Littlewood conjecture is trivially satisfied. Indeed, if α is not badly approximable, there is a sequence q_n such that

$$q_n \|q_n \alpha\| = q_n^2 \left| \alpha - \frac{p_n}{q_n} \right| \to 0.$$

Brutally estimating $\|q_n \beta\| \leq 1/2$, we find that

$$q_n \|q_n \alpha\| \|q_n \beta\| \to 0,$$

so that the pair (α, β) satisfies the Littlewood conjecture. Hence, this problem naturally lives on a set of measure zero, namely Bad $\times$ Bad.

The key tool in proving Theorem 6.7 is a result on the discrepancy of certain sequences, which holds true for almost all α with respect to a certain measure introduced by Kaufman [32].

Kaufman's measure μ_M is a measure supported on the set of real numbers with partial quotients bounded above by M. To be explicit, for each real number $\alpha \in [0, 1)$, let

$$\alpha = [a_1, a_2, \dots] = \cfrac{1}{a_1 + \cfrac{1}{a_2 + \cfrac{1}{\ddots}}}$$

be the simple continued fraction expansion of α. For $M \geq 3$, let

$$F_M = \{\alpha \in [0, 1) : a_i(\alpha) \leq M \text{ for all } i \in \mathbb{N}\}. \tag{6.8}$$

Recall that the set of badly approximable numbers consists exactly of the numbers for which the partial quotients form a bounded sequence, so that

$$\text{Bad} = \bigcup_{M=1}^{\infty} F_M.$$

Kaufman proved that the set F_M supports a measure μ_M satisfying a number of nice properties. For our purposes, we need the following two properties.

(i) For any $s < \dim_H(F_M)$, there are positive constants $c, l > 0$ such that for any interval $I \subseteq [0, 1)$ of length $|I| \leq l$,

$$\mu_M(I) \leq c \, |I|^s.$$

(ii) For any M, there are positive constants $c, \eta > 0$ such that the Fourier transform $\hat{\mu}_M$ of the Kaufman measure μ_M satisfies

$$\hat{\mu}_M(u) \leq c \, |u|^{-\eta}.$$

The first property allows us to connect the Kaufman measure with the Hausdorff dimension of the set F_M via the mass distribution principle of Lemma 3.3. The second property provides a positive lower bound on the Fourier dimension of the set F_M, but for our purposes the property is used only in computations.

The second key tool is the notion of discrepancy from the theory of uniform distribution. The discrepancy of a sequence in $[0, 1)$ measures how uniformly distributed a sequence is in the interval. Specifically, the discrepancy of the sequence (x_n) is defined as

$$D_N(x_n) = \sup_{I \subseteq [0,1]} \left| \sum_{n=1}^{N} \chi_I(x_n) - N |I| \right|,$$

where I is an interval and χ_I is the corresponding characteristic function. A sequence (x_n) is uniformly distributed if $D_N(x_n) = o(N)$.

Our key result is the following discrepancy estimate, which implies Theorem 6.7

Theorem 6.8 *Let μ_M be a Kaufman measure and assume that for positive integers $u < v$ we have*

$$\sum_{n,m=u}^{v} |a_n - a_m|^{-\eta} \ll \frac{1}{\log v} \sum_{n=u}^{v} \psi_n$$

where (ψ_n) is a sequence of non-negative numbers and $\eta > 0$ is the constant from property (ii) of the Kaufman measure. Then for μ_M-almost every $x \in [0, 1]$ we have

$$D_N(a_n x) \ll (N \log(N)^2 + \Psi_N)^{1/2} \log(N \log(N)^2 + \Psi_N)^{3/2+\varepsilon} + \max_{n \leq N} \psi_n$$

where $\Psi_N = \psi_1 + \cdots + \psi_N$.

We will need a probabilistic lemma which can be found in [26].

Lemma 6.9 *Let (X, μ) be a measure space with $\mu(X) < \infty$. Let $F(n, m, x)$, $n, m \geq 0$ be μ-measurable functions and let ϕ_n be a sequence of real numbers such that $|F(n-1, n, x)| \leq \phi_n$ for $n \in \mathbb{N}$. Let $\Phi_N = \phi_1 + \cdots + \phi_N$ and assume that $\Phi_N \to \infty$. Suppose that for $0 \leq u < v$ we have*

$$\int_X |F(u, v, x)|^2 \, d\mu \ll \sum_{n=u}^{v} \phi_n.$$

Then for μ-almost all x, we have

$$F(0, N, x) \ll \Phi_N^{1/2} \log(\Phi_N)^{3/2+\varepsilon} + \max_{n \leq N} \phi_n.$$

We will also need the classical Erdős–Turán inequality which can be found in [40].

Theorem 6.10 *For any positive integer K and any sequence $(x_n) \subseteq [0, 1)$,*

$$D_N(x_n) \leq \frac{N}{K+1} + 3 \sum_{k=1}^{K} \frac{1}{k} \left| \sum_{n=1}^{N} e(kx_n) \right|,$$

where as usual $e(x) = \exp(2\pi i x)$.

Proof of Theorem 6.8. Suppose $M \geq 3$ and for integers $0 \leq u < v$ let

$$F(u, v, x) = \sum_{h=1}^{v} \frac{1}{h} \left| \sum_{n=u}^{v} e(ha_n x) \right|.$$

Theorem 6.10 with $K = N$ tells us that

$$D_N(a_n x) \ll F(0, N, x).$$

Integrating with respect to $d\mu_M(x)$ and applying the Cauchy–Schwarz inequality gives

$$\int |F(u, v, x)|^2 \, d\mu_M \leq \sum_{h,k=1}^{v} \frac{1}{hk} \int \left| \sum_{n=u}^{v} e(ha_n x) \right|^2 d\mu_M$$

$$= \sum_{h,k=1}^{v} \frac{1}{hk} \left(v - u + 1 + \sum_{\substack{n,m=u \\ n \neq m}}^{v} \hat{\mu}_M(h(a_n - a_m)) \right).$$

Finally using property (ii) of the Kaufman measure we have

$$\int |F(u, v, x)|^2 \, d\mu_M \ll \sum_{h,k=1}^{v} \frac{1}{hk} \left(v - u + 1 + h^{-\eta} \sum_{\substack{n,m=u \\ n \neq m}}^{v} |a_n - a_m|^{-\eta} \right)$$

$$\ll \sum_{n=u}^{v} [\log(n)^2 + \psi_n].$$

Since $F(n - 1, n, x) \ll \log(n)^2 + \psi_n$ for all $n \geq 1$, the theorem then follows from Lemma 6.9.

For sequences which grow sufficiently rapidly, the theorem has a corollary with a much cleaner statement. We will say that an increasing sequence of positive integers (a_n) is *lacunary* if there is a $c > 1$ such that for any n, $a_{n+1}/a_n > c$. Applying this inductively, we see that the sequence must grow at least as fast as some geometric sequence.

Corollary 6.11 *Let $\nu > 0$, let μ be a Kaufman measure and (a_n) a lacunary sequence of integers. For μ-almost every $x \in [0, 1]$ we have $D_N(a_n x) \ll N^{1/2} (\log N)^{5/2+\nu}$.*

Proof We apply again Theorem 6.8. Using lacunarity of the sequence (a_n), we see that

$$\sum_{n,m=1}^{\infty} |a_n - a_m|^{-\eta} < \infty.$$

Consequently, we can absorb all occurrences of Ψ_N as well as the final term $\max_{n \leq N} \psi_n$ in the discrepancy estimate of Theorem 6.8 into the implied constant. It follows that

$$D_N(a_n x) \ll (N \log(N)^2)^{1/2} \log(N \log(N)^2)^{3/2+\varepsilon} \ll N^{1/2} (\log N)^{5/2+\nu}$$

for μ-almost every x, where ν can be made as small as desired by picking ε small enough.

Proof of Theorem 6.7. Let G denote the set of numbers $\beta \in \text{Bad}$ for which there is an i, such that

$$\liminf q \|q\alpha_i\| \|q\beta\| > 0. \tag{6.9}$$

Suppose, contrary to what we are to prove, that $\dim_H G < 1$. Pick an $M \geq 3$ such that $\dim_H F_M > \dim_H G$ (this can be done in light of Jarník's theorem). Let $\mu = \mu_M$ denote the Kaufman measure on F_M.

Consider first one of the α_i, and let (q_k) denote the sequence of denominators of convergents in the simple continued fraction expansion of α_i. In the following, we will use the various parts of Proposition 1.4 many times to deduce results about this sequence and its relation to α_i.

The sequence q_k is lacunary. Hence, by Corollary 6.11, for μ–almost every x,

$$D_N(q_n x) \ll N^{1/2} (\log N)^{5/2+\nu}.$$

Let $\psi(N) = N^{-1/2+\epsilon}$ for some $\epsilon > 0$ and consider the interval

$$I_N = [-\psi(N), \psi(N)].$$

By the definition of discrepancy, for every $\gamma \in [0, 1]$ and μ-almost every β

$$|\#\{k \leq N : \{q_k \beta\} \in I_N\} - 2N\psi(N)| \ll N^{1/2} (\log N)^{5/2+\nu}.$$

Hence,

$$\#\{k \leq N : \{q_k \beta\} \in I_N\} \geq 2N\psi(N) - KN^{1/2} (\log N)^{5/2+\nu}$$
$$= 2N^{1/2+\epsilon} - KN^{1/2} (\log N)^{5/2+\nu},$$

where $K > 0$ is the implied constant from Corollary 6.11. Next let N_h denote the increasing sequences defined by

$$N_h = \min \left\{ N \in \mathbb{N} : \#\{k \le N : \{q_k \beta\} \in I_N\} = h \right\}.$$

Since each q_{N_h} is a denominator of a convergent to α_i,

$$q_{N_h} \left\| q_{N_h} \alpha_i \right\| \le 1.$$

Hence,

$$q_{N_h} \left\| q_{N_h} \alpha_i \right\| \left\| q_{N_h} \beta \right\| \le \left\| q_{N_h} \beta \right\| \le \left(N_h^{\gamma} \right)^{-1/2+\epsilon}.$$

This establishes our claim and shows that the exceptional set $E_i \subseteq F_M$ for which (6.9) holds has $\mu(E_i) = 0$.

To conclude, let E be the set of $\beta \in F_M$ for which there is an i or a j such that either (6.4) or (6.5) is not satisfied. Then,

$$E = \bigcup_i E_i \cup \bigcup_j E_j',$$

and therefore $\mu(E) = 0$.

Finally $\mu(G)$ is maximal, so consider the trace measure $\tilde{\mu}$ of μ on G, defined by $\tilde{\mu}(X) = \mu(X \cap G)$. It follows from property (i) of Kaufman's measures that μ is a mass distribution on $[0, 1)$, and since G is full, $\tilde{\mu}$ inherits the decay property of (i) from μ. By the mass distribution principle it then follows that $\dim_H(G) = \dim_H(F_M) > \dim_H(G)$, which contradicts our original assumption. Therefore we conclude that $\dim_H(G) = 1$.

The proof of Theorem 6.7 in fact tells us that something stronger than the Littlewood conjecture holds for the pairs (α_i, β). Indeed, we can work a little more with the inequalities obtained and get a speed of convergence along the sequence (q_{N_h}). Further results using the full force of the uniform distribution of the sequence $(q_n \beta)$ can be found in the original paper [28], where we also prove similar results for the related p-adic and mixed Littlewood conjectures. However, it is beyond the scope of these notes to discuss these topics, and for the clarity of the exposition we have restricted ourselves to results on the original conjecture.

7 Concluding Remarks

These notes are far from being a complete description of the state-of-the-art in metric Diophantine approximation. Recent developments in metric Diophantine approximation on manifolds has barely been touched upon, and the relation with homogeneous dynamics which has led to spectacular advances in the theory has only been superficially described. The selection of results reflect the tastes and expertises of the author, and much is left out.

Nonetheless, it is hoped that the reader has caught a glimpse of the richness and beauty of the metric theory of Diophantine approximation and has acquired a taste for more. Certainly, there are problems and literature enough to last a lifetime of research.

Acknowledgements I acknowledge the support of the Danish Natural Science Research Council. I am grateful to the organisers and participants of the Summer School 'Diophantine Analysis' in Würzburg 2014. Finally, I warmly thank Kalle Leppälä, Steffen Højris Pedersen and Morten Hein Tiljeset for their comments on various drafts of this manuscript.

References

1. J. An, Badziahin-Pollington-Velani's theorem and Schmidt's game. Bull. Lond. Math. Soc. **45**(4), 721–733 (2013)
2. D. Badziahin, S. Velani, Badly approximable points on planar curves and a problem of Davenport. Math. Ann. **359**(3–4), 969–1023 (2014)
3. D. Badziahin, A. Pollington, S. Velani, On a problem in simultaneous Diophantine approximation: Schmidt's conjecture. Ann. of Math. (2) **174**(3), 1837–1883 (2011)
4. V. Beresnevich, On approximation of real numbers by real algebraic numbers. Acta Arith. **90**(2), 97–112 (1999)
5. V. Beresnevich, Badly approximable points on manifolds. Invent. Math. **202**(3), 1199–1240 (2015)
6. V. Beresnevich, S. Velani, Schmidt's theorem, Hausdorff measures, and slicing. Int. Math. Res. Not. **24** (2006). (Art. ID 48794)
7. V. Beresnevich, S. Velani, A mass transference principle and the Duffin-Schaeffer conjecture for Hausdorff measures. Ann. Math. (2) **164**(3), 971–992 (2006)
8. V. Beresnevich, D. Dickinson, S. Velani, Measure theoretic laws for lim sup sets. Mem. Amer. Math. Soc. **179**(846) (2006). (x+91)
9. V.I. Bernik, The exact order of approximating zero by values of integral polynomials. Acta Arith. **53**(1), 17–28 (1989)
10. A.S. Besicovitch, Sets of fractional dimension (IV); on rational approximation to real numbers. J. Lond. Math. Soc. **9**, 126–131 (1934)
11. P. Billingsley, *Ergodic Theory and Information* (Robert E. Krieger Publishing Co., Huntington, 1978)
12. É. Borel, Les probabilités dénombrables et leurs applications arithmétiques. Rend. Circ. Math. Palermo **27**, 247–271 (1909)
13. Y. Bugeaud, *Approximation by Algebraic Numbers* (Cambridge University Press, Cambridge, 2004)
14. Y. Bugeaud, M.M. Dodson, S. Kristensen, Zero-infinity laws in Diophantine approximation. Q. J. Math. **56**(3), 311–320 (2005)
15. C. Carathéodory, *Über das lineare Mass von Punktmengen, eine Verallgemeinerung des Längenbegriffs*, Gött. Nachr. 404–426 (2014)
16. J.W.S. Cassels, *An Introduction to Diophantine Approximation* (Cambridge University Press, New York, 1957)
17. J.W.S. Cassels, *An Introduction to the Geometry of Numbers* (Springer-Verlag, Berlin, 1997)
18. L.G.P. Dirichlet, *Verallgemeinerung eines Satzes aus der Lehre von den Kettenbrüchen nebst einige Anwendungen auf die Theorie der Zahlen*, S.-B. Preuss. Akad Wiss. 93–95 (1842)

19. M.M. Dodson, *Geometric and probabilistic ideas in the metric theory of Diophantine approximations*. Uspekhi Mat. Nauk **48**(5)(293), 77–106 (1993)

20. M.M. Dodson, B.P. Rynne, J.A.G. Vickers, Diophantine approximation and a lower bound for Hausdorff dimension. Mathematika **37**(1), 59–73 (1990)

21. R.J. Duffin, A.C. Schaeffer, Khintchine's problem in metric Diophantine approximation. Duke Math. J. **8**, 243–255 (1941)

22. M. Einsiedler, T. Ward, *Ergodic Theory with a View Towards Number Theory* (Springer, London, 2011)

23. M. Einsiedler, A. Katok, E. Lindenstrauss, *Invariant measures and the set of exceptions to Littlewood's conjecture*. Ann. Math. (2) **164**(2), 513–560 (2006)

24. K. Falconer, *Fractal Geometry, Mathematical Foundations and Applications* (Wiley, Hoboken, 2003)

25. O. Frostman, Potentiel d'équilibre et capacité des ensembles avec quelques applications à la théorie des fonctions. Meddel. Lunds Univ. Math. Sem **3**, 1–118 (1935)

26. G. Harman, *Metric Number Theory* (The Clarendon Press, Oxford University Press, New York, 1998)

27. F. Hausdorff, Dimension und äusseres Mass. Math. Ann. **79**, 157–179 (1919)

28. A. Haynes, J.L. Jensen, S. Kristensen, Metrical musings on Littlewood and friends. Proc. Amer. Math. Soc. **142**(2), 457–466 (2014)

29. V. Jarník, *Zur metrischen Theorie der diophantischen Approximationen*, Prace Mat.-Fiz. 91–106 (1928–1929)

30. V. Jarník, Diophantischen Approximationen und Hausdorffsches Mass. Mat. Sbornik **36**, 91–106 (1929)

31. J.-P. Kahane, R. Salem, *Ensembles parfaits et séries trigonométriques* (Hermann, Paris, 1994)

32. R. Kaufman, Continued fractions and Fourier transforms. Mathematika **27**(2), 262–267 (1980)

33. D. Kleinbock, B. Weiss, Badly approximable vectors on fractals. Israel J. Math. **149**, 137–170 (2005)

34. D. Kleinbock, E. Lindenstrauss, B. Weiss, *On fractal measures and Diophantine approximation*, Selecta Math. (N.S.) **10**(4), 479–523 (2004)

35. S. Kristensen, *Approximating numbers with missing digits by algebraic numbers*. Proc. Edinb. Math. Soc. (2) **49**(3), 657–666 (2006)

36. S. Kristensen, R. Thorn, S. Velani, Diophantine approximation and badly approximable sets. Adv. Math. **203**(1), 132–169 (2006)

37. J. Levesley, C. Salp, S. Velani, On a problem of K. Mahler: Diophantine approximation and Cantor sets. Math. Ann. **338**(1), 97–118 (2007)

38. P. Mattila, *Geometry of Sets and Measures in Euclidean Spaces* (Cambridge University Press, Cambridge, 1995)

39. H. Minkowski, Geometrie der Zahlen, Leipzig and Berlin (1896)

40. H. Montgomery, *Ten Lectures on the Interface Between Analytic Number Theory and Harmonic Analysis* (American Mathematical Society, Providence, 1994)

41. G.K. Pedersen, *Analysis Now* (Springer, New York, 1989)

42. A.D. Pollington, S. Velani, On a problem in simultaneous Diophantine approximation: Littlewood's conjecture. Acta Math. **185**(2), 287–306 (2000)

43. E.T. Poulsen, A simplex with dense extreme points. Ann. Inst. Fourier. Grenoble **11**, 83–87 (1961)

44. W.M. Schmidt, *Über die Normalität von Zahlen zu verschiedenen Basen*. Acta Arith. **7**, 299–309 (1961/1962)

45. W.M. Schmidt, On badly approximable numbers and certain games. Trans. Am. Math. Soc. **123**, 178–199 (1966)

46. C. Series, *The modular surface and continued fractions*. J. London Math. Soc. (2) **31**(1), 69–80 (1985)

47. V.G. Sprindžuk, *Metric Theory of Diophantine Approximations* (Wiley, London, 1979)

48. E. Wirsing, Approximation mit Algebraischen Zahlen beschränkten Grades. J. Reine Angew. Math. **206**, 67–77 (1961)

49. A. Ya, *Khintchine* (Dover Publications Inc, Mineola, NY, Continued fractions, 1997)

A Geometric Face of Diophantine Analysis

Tapani Matala-aho

2000 Mathematics Subject Classification 11H06 · 11J25 · 11J81 · 11J82

1 Targets

Our first target is to prove some Diophantine inequalities over real numbers by using Minkowski's first convex body theorem. Also some generalizations over complex numbers are discussed.

Next we give an elementary proof of Siegel's lemma over rational numbers. Then we present without proof a version of Siegel's lemma over an arbitrary imaginary quadratic field which slightly improves the existing versions.

Our following target is to give a proof to the equality of heights of a rational subspace and its orthogonal complement. The proof is based on Grassmann algebra. Therefore, quite an amount of basics of exterior algebras will be presented in the framework of rational subspaces. An important tool will be the primitive Grassmann coordinate vector of the corresponding rational subspace. Lastly, we are ready to prove the Bombieri–Vaaler [3] version of Siegel's lemma.

2 Convex Bodies and Lattices

In this section we shall present notations and basics of convex bodies and lattices. For a more comprehensive presentation we refer to Steuding [14], Sects. 8.1–8.3, Cassels [5] or Schmidt [10].

T. Matala-aho (✉)
Matemaattisten Tieteiden Laitos, Oulun Yliopisto, PL 3000, 90014 Oulu, Finland
e-mail: tapani.matala-aho@oulu.fi

© Springer International Publishing AG 2016
J. Steuding (ed.), *Diophantine Analysis*, Trends in Mathematics,
DOI 10.1007/978-3-319-48817-2_3

129

Throughout these notes the column vectors are denoted by $\bar{x} = (x_1, \ldots, x_n)^t \in \mathbb{R}^n$, where the coordinates belong to a ring (or a field) R specified later.

For the different norms of the vector $\bar{x} = (x_1, \ldots, x_n)^t \in \mathbb{C}^n$ we shall use the notations

$$\|\bar{x}\|_\infty = \max_{k=1,\ldots,n} |x_k|, \quad \|\bar{x}\|_1 = \sum_{k=1}^n |x_k|, \quad \|\bar{x}\|_2 = \|\bar{x}\| = \left(\sum_{k=1}^n |x_k|^2\right)^{1/2},$$

where the first is so-called maximum norm and the last is the usual Euclidean norm.

2.1 Convex Bodies

A non-empty subset $\mathcal{C} \subseteq \mathbb{R}^n$ is *convex*, if for any pair of points $\bar{a}, \bar{b} \in \mathcal{C}$ holds

$$\{s\bar{a} + (1-s)\bar{b}|\ 0 \leq s \leq 1\} \subseteq \mathcal{C}.$$

A bounded convex subset $\mathcal{C} \subseteq \mathbb{R}^n$ is called a *convex body*. In these notes we don't expect that a convex body is necessarily closed. A subset $\mathcal{C}$ is *central symmetric* (symmetric wrt origin) if $\mathcal{C} = -\mathcal{C}$.

Example 2.1 Let $\lambda \in \mathbb{R}^+$ and assume that $\mathcal{C}$ is a central symmetric convex body. Then the dilation

$$\lambda\mathcal{C} := \{\lambda\bar{a}|\ \bar{a} \in \mathcal{C}\}$$

is also a central symmetric convex body.

By a volume $V(\mathcal{C})$ of a subset $\mathcal{C} \subseteq \mathbb{R}^n$ we mean the Riemann (or Lebesgue) integral, if it exists.

2.2 Rings and Modules

In the following, a ring, say $(R, \times, +)$, is a non-empty set R with ring product $\times$ and addition $+$. The ring is not necessarily commutative but it has zero and identity elements $0, 1 \in R, 0 \neq 1$.

Let R be a commutative ring. Then $(M, +, \cdot)$ is an *R-module*, if $(M, +)$ is an Abelian group and the scalar product $\cdot : R \times M \to M$ satisfies the following axioms

$$1 \cdot m = m, \ 1 \in R; \quad (rs) \cdot m = r \cdot (s \cdot m);$$
$$(r+s) \cdot m = r \cdot m + s \cdot m; \ r \cdot (m+n) = r \cdot m + r \cdot n \quad \forall r, s \in R, \ m, n \in M.$$

The elements of R are called scalars. Let M be an R-module, S a subring of R and $B \subseteq M$, then

$$\langle B \rangle_S = \bigcap_{B \subseteq N} N, \quad N \text{ is a } S\text{-submodule of } M,$$

is called *a linear hull* over S generated by B. In particular,

$$\langle m_1, \ldots, m_k \rangle = \langle m_1, \ldots, m_k \rangle_S := Sm_1 + \cdots + Sm_k$$

is called *a linear hull* over S generated by $m_1, \ldots, m_k \in M$.

Example 2.2 If K is a field, then the K-vector space $M = K^n$ is a K-module. We may write

$$K^n = \langle \bar{e}_1, \ldots, \bar{e}_n \rangle_K = K\bar{e}_1 + \cdots + K\bar{e}_n,$$

where $\bar{e}_1 = (1, 0, \ldots, 0)^t, \ldots, \bar{e}_n = (0, \ldots, 0, 1)^t$ denote the standard (column) base vectors.

2.3 Lattices

In these lectures we consider lattices which are free additive subgroups of $(\mathbb{R}^n, +)$. Let $n \in \mathbb{Z}^+$ and let $\bar{l}_1, \ldots, \bar{l}_r \in \mathbb{R}^n$ be linearly independent over $\mathbb{R}$, then the linear hull

$$\Lambda = \langle \bar{l}_1, \ldots, \bar{l}_r \rangle_{\mathbb{Z}} = \mathbb{Z}\bar{l}_1 + \cdots + \mathbb{Z}\bar{l}_r \subseteq \mathbb{R}^n$$

over $\mathbb{Z}$ forms a lattice. The set $\{\bar{l}_1, \ldots, \bar{l}_r\}$ is called a base of Λ with rank $\Lambda = r$. If rank $\Lambda = n$, then Λ is called a full lattice. The determinant of Λ is defined by

$$\det(\Lambda) := \sqrt{\det[\bar{l}_i \cdot \bar{l}_j]_{1 \leq i,j \leq r}} = \sqrt{\det(L^t L)}, \quad L = [\bar{l}_1, \ldots, \bar{l}_r].$$

where the columns of the matrix L are the base vectors $\bar{l}_1, \ldots, \bar{l}_r$ of Λ. In particular, for the full lattice we have

$$\det(\Lambda) = |\det L| = \left| \det[\bar{l}_1, \ldots, \bar{l}_n] \right|.$$

Example 2.3 The integer lattice

$$\mathbb{Z}^n = \mathbb{Z}\bar{e}_1 + \cdots + \mathbb{Z}\bar{e}_n$$

has determinant $\det(\Lambda) = 1$.

3 Minkowski's Convex Body Theorems

3.1 The First Minkowski's Convex Body Theorem

Theorem 3.1 *The first Minkowski's convex body theorem. Let $n \in \mathbb{Z}^+$. Assume that $\Lambda \subseteq \mathbb{R}^n$ is a lattice with rank $\Lambda = n$ and $\mathcal{C} \subseteq \mathbb{R}^n$ is a central symmetric convex body with*

$$\begin{cases} V(\mathcal{C}) > 2^n \det \Lambda \quad or \\ V(\mathcal{C}) \geq 2^n \det \Lambda, \quad if \quad \mathcal{C} \quad is\ compact. \end{cases}$$

Then, there exists a non-zero lattice point in $\mathcal{C}$. In fact, then $\#\mathcal{C} \cap \Lambda \geq 3$.

3.2 The Second Minkowski's Convex Body Theorem

3.2.1 Successive Minima

Definition 3.2 Let $n \in \mathbb{Z}^+$ and let $\mathcal{C}$ be a non-empty subset of $\mathbb{R}^n$. The successive minima $\lambda_1, \ldots \lambda_n$ of $\mathcal{C}$ with respect to a lattice $\Lambda \subseteq \mathbb{R}^n$ are given by

$$\lambda_j = \inf\{\lambda > 0 |\ \text{rank}\langle (\lambda\mathcal{C}) \cap \Lambda \rangle_{\mathbb{Z}} \geq j\}.$$

Lemma 3.3
$$0 < \lambda_1 \leq \cdots \leq \lambda_n < \infty.$$

Example 3.4 For $n \in \mathbb{Z}^+$ let

$$\mathcal{B}_n := \{\overline{x} \in \mathbb{R}^n |\ \|\overline{x}\|_2 \leq 1\}$$

denote the unit ball in $\mathbb{R}^n$ and let $\lambda_1, \ldots \lambda_n$ be the successive minima of $\mathcal{B}_n$ with respect to a lattice $\Lambda \subseteq \mathbb{R}^n$. Then, by the definition of successive minima there exist n linearly independent integer vectors $\overline{x}_1, \ldots, \overline{x}_n \in \Lambda \setminus \{\overline{0}\}$ such that

$$\|\overline{x}_1\|_2 \leq \lambda_1, \ldots, \|\overline{x}_n\|_2 \leq \lambda_n. \tag{3.1}$$

Example 3.5 For $n \in \mathbb{Z}^+$ let

$$\square_n := \{\overline{x} \in \mathbb{R}^n |\ \|\overline{x}\|_\infty \leq 1\}$$

denote the n-dimensional "unit" cube in $\mathbb{R}^n$. Let $\lambda_1, \ldots \lambda_n$ be the successive minima of $\square_n$ with respect to a lattice $\Lambda \subseteq \mathbb{R}^n$. Then, there exist n linearly independent integer vectors $\overline{x}_1, \ldots, \overline{x}_n \in \Lambda \setminus \{\overline{0}\}$ such that

$$\|\overline{x}_1\|_\infty \leq \lambda_1, \ldots, \|\overline{x}_n\|_\infty \leq \lambda_n. \tag{3.2}$$

3.2.2 The Second Minkowski's Convex Body Theorem

Theorem 3.6 *Let $n \in \mathbb{Z}^+$. Assume that $\Lambda \subseteq \mathbb{R}^n$ is a lattice with rank $\Lambda = n$ and $C \subseteq \mathbb{R}^n$ is a central symmetric convex body. Then*

$$\frac{2^n}{n!} \det \Lambda \leq \lambda_1 \cdots \lambda_n V(C) \leq 2^n \det \Lambda.$$

4 Diophantine Inequalities Over $\mathbb{R}$

In this chapter we present some selected Diophantine inequalities over $\mathbb{R}$. For further studies we recommend, Cassels [5, 11, 12], Shidlovskii [12] and Steuding [14].

We start from Theorem 4.1, which is a typical example from the theory of linear forms.

Theorem 4.1 *Let*

$$\alpha_1, \ldots, \alpha_m \in \mathbb{R}$$

and

$$h_1, \ldots, h_m \in \mathbb{Z}^+$$

be given. Then there exist $p, q_1, \ldots, q_m \in \mathbb{Z}$ with a $q_k \neq 0$, satisfying

$$|q_i| \leq h_i, \quad i = 1, \ldots, m,$$

and

$$|p + q_1\alpha_1 + \cdots + q_m\alpha_m| < \frac{1}{h_1 \cdots h_m}.$$

Proof Write

$$L_0\overline{x} := x_0 + \alpha_1 x_1 + \cdots + \alpha_m x_m,$$

$$L_k\overline{x} := x_k, \quad k = 1, \ldots, m$$

and put

$$|L_0\overline{x}| < \frac{1}{h} := \frac{1}{h_1 \cdots h_m}, \tag{4.1}$$

$$|L_k\overline{x}| \leq h_k + 1/2, \quad k = 1, \ldots, m. \tag{4.2}$$

Then

$$(L_0, L_1, \ldots, L_m) : \mathbb{Z}^{m+1} \to \mathbb{R}^{m+1}$$

defines a full lattice

$$\Lambda := \mathbb{Z}\bar{\ell}_0 + \mathbb{Z}\bar{\ell}_1 + \cdots + \mathbb{Z}\bar{\ell}_m =$$
$$\mathbb{Z}(1, 0, \ldots, 0)^t + \mathbb{Z}(\alpha_1, 1, 0, \ldots, 0)^t + \cdots + \mathbb{Z}(\alpha_m, 0, \ldots, 1)^t \subseteq \mathbb{R}^{m+1}$$

with determinant

$$\det \Lambda = \left| \det(\bar{\ell}_0\, \bar{\ell}_1 \ldots \bar{\ell}_m) \right| = \begin{vmatrix} 1 & \alpha_1 & \alpha_2 & \ldots & \alpha_m \\ 0 & 1 & 0 & \ldots & 0 \\ 0 & 0 & 1 & \ldots & 0 \\ . & . & . & \ldots & . \\ 0 & 0 & 0 & \ldots & 1 \end{vmatrix} = 1.$$

Now the conditions (4.1) and (4.2) determine a convex set

$$\mathcal{C} := \{(y_0, \ldots, y_m)^t \mid |y_0| < 1/h, \ |y_k| \leq h_k + 1/2\} \subseteq \mathbb{R}^{m+1}$$

with volume

$$V(\mathcal{C}) = \frac{2^{m+1}(h_1 + 1/2) \cdots (h_m + 1/2)}{h_1 \cdots h_m} > 2^{m+1}.$$

By the first Minkowski's convex body there exists a non-zero vector

$$\bar{0} \neq \bar{y} = (p + q_1\alpha_1 + \cdots + q_m\alpha_m, q_1, \ldots, q_m)^t \in \mathcal{C} \cap \Lambda. \tag{4.3}$$

such that

$$|q_i| \leq h_i, \quad \forall i = 1, \ldots, m,$$

and

$$|p + q_1\alpha_1 + \cdots + q_m\alpha_m| < \frac{1}{h_1 \cdots h_m}, \tag{4.4}$$

where by (4.3) we have

$$(p, q_1, \ldots, q_m)^t \in \mathbb{Z}^{m+1} \setminus \{\bar{0}\}.$$

Finally, if all $q_1 = \cdots = q_m = 0$ in (4.4), then also $p = 0$. A contradiction. □

4.1 Primitive Vector

An integer vector

$$(r_0, r_1, \ldots, r_m)^t \in \mathbb{Z}^{m+1}$$

is *primitive*, if the greatest common divisor satisfies

$$\gcd(r_0, r_1, \ldots, r_m) = 1.$$

Let

$$\gcd(p, q_1, \ldots, q_m) = d \in \mathbb{Z}_{\geq 2}$$

and suppose the integer vector

$$(p, q_1, \ldots, q_m)^t = d(s, r_1, \ldots, r_m)^t, \quad (s, r_1, \ldots, r_m)^t \in \mathbb{Z}^{m+1},$$

satisfies the estimate

$$|p + q_1\alpha_1 + \cdots + q_m\alpha_m| < \frac{1}{h}.$$

Then

$$|s + r_1\alpha_1 + \cdots + r_m\alpha_m| < \frac{1}{dh} < \frac{1}{h}.$$

Thus we have also a primitive solution

$$(s, r_1, \ldots, r_m)^t \in \mathbb{Z}^{m+1}$$

for equation

$$|p + q_1\alpha_1 + \cdots + q_m\alpha_m| < \frac{1}{h}.$$

Theorem 4.2 *Let*

$$\alpha_1, \ldots, \alpha_m \in \mathbb{R}$$

and

$$h_1, \ldots, h_m \in \mathbb{Z}^+$$

be given. Then there exists a primitive vector

$$(p, q_1, \ldots, q_m)^t \in \mathbb{Z}^{m+1} \setminus \{\bar{0}\}$$

with a $q_k \neq 0$, satisfying

$$|q_i| \leq h_i, \quad \forall i = 1, \ldots, m,$$

and

$$|p + q_1\alpha_1 + \cdots + q_m\alpha_m| < \frac{1}{h_1 \cdots h_m}.$$

4.2 Infiniteness of Primitive Solutions

Theorem 4.3 *Let*

$$1, \alpha_1, \ldots, \alpha_m \in \mathbb{R}$$

be linearly independent over $\mathbb{Q}$. *Then there exist infinitely many primitive vectors*

$$\overline{v}_k = (p_k, q_{1,k}, \ldots, q_{m,k})^t \in \mathbb{Z}^{m+1} \setminus \{\overline{0}\} \tag{4.5}$$

with

$$h_{i,k} := \max\{1, |q_{i,k}|\}, \quad \forall i = 1, \ldots, m,$$

satisfying

$$|p_k + q_{1,k}\alpha_1 + \cdots + q_{m,k}\alpha_m| < \frac{1}{h_{1,k} \cdots h_{m,k}} := \frac{1}{h_k}. \tag{4.6}$$

Proof Suppose on the contrary, that there exist only finitely many primitive solutions for (4.6). Then by the linear independence and assumption (4.5) there exists a minimum

$$\min |p_k + q_{1,k}\alpha_1 + \cdots + q_{m,k}\alpha_m| := \frac{1}{R} > 0. \tag{4.7}$$

Choose then

$$\hat{h}_i \in \mathbb{Z}^+, \quad \hat{h} := \hat{h}_1 \cdots \hat{h}_m, \quad \frac{1}{\hat{h}} \le \frac{1}{R}.$$

Now by Theorem 4.2 there exists a primitive solution

$$(\hat{p}, \hat{q}_1, \ldots, \hat{q}_m)^t \in \mathbb{Z}^{m+1} \setminus \{\overline{0}\}$$

with

$$\max\{1, |\hat{q}_i|\} \le \hat{h}_i, \quad \forall i = 1, \ldots, m,$$

satisfying

$$|\hat{p} + \hat{q}_1\alpha_1 + \cdots + \hat{q}_m\alpha_m| < \frac{1}{\hat{h}} \le \frac{1}{R}.$$

which contradicts (4.7). □

4.3   Corollaries

Theorem 4.4 *Let*

$$1, \alpha_1, \ldots, \alpha_m \in \mathbb{R}$$

be linearly independent over $\mathbb{Q}$. If there exist positive constants $c, \omega \in \mathbb{R}^+$ such that

$$|\beta_0 + \beta_1 \alpha_1 + \cdots + \beta_m \alpha_m| \geq \frac{c}{(h_1 \cdots h_m)^\omega} \tag{4.8}$$

holds for all

$$(\beta_0, \beta_1, \ldots, \beta_m) \in \mathbb{Z}^{m+1} \setminus \{\bar{0}\}, \quad h_k = \max\{1, |\beta_k|\},$$

then

$$\omega \geq 1.$$

Proof Assume on the contrary that

$$\omega < 1.$$

By Theorem 4.3 there exists an infinity of primitive vectors satisfying

$$|p_k + q_{1,k} \alpha_1 + \cdots + q_{m,k} \alpha_m| < \frac{1}{h_{1,k} \cdots h_{m,k}}$$

and by (4.8) we have

$$\frac{c}{(h_{1,k} \cdots h_{m,k})^\omega} \leq |p_k + q_{1,k} \alpha_1 + \cdots + q_{m,k} \alpha_m| < \frac{1}{h_{1,k} \cdots h_{m,k}}. \tag{4.9}$$

Choose now $h_{1,k}, \ldots, h_{m,k}$ such that

$$\log(h_{1,k} \cdots h_{m,k}) \geq \frac{\log(1/c)}{1 - \omega}.$$

A contradiction with (4.9). □

Usually, in the existing literature, the above results are only given in terms of

$$H_k := \max_{i=1,\ldots,m} |q_{i,k}|.$$

Theorem 4.5 *Let*

$$1, \alpha_1, \ldots, \alpha_m \in \mathbb{R}$$

be linearly independent over $\mathbb{Q}$. *Then there exist infinitely many primitive vectors*

$$\bar{v}_k = (p_k, q_{1,k}, \ldots, q_{m,k})^t \in \mathbb{Z}^{m+1} \setminus \{\bar{0}\}$$

satisfying

$$|p_k + q_{1,k}\alpha_1 + \cdots + q_{m,k}\alpha_m| < \frac{1}{H_k^m}. \tag{4.10}$$

One may wonder, if the exponent in (4.10) could be improved? Theorem 4.6 shows that the upper bound in (4.10) is best possible up to a constant factor for an arbitrary m-tuple $(\alpha_1, \ldots, \alpha_m)$ of real numbers.

Theorem 4.6 *Let* $\alpha = \alpha_0$ *be an algebraic integer of degree* $\deg_{\mathbb{Q}} \alpha = m + 1$ *and* $\alpha_i = \sigma_i(\alpha)$, $i = 0, 1, \ldots, m$, *where* σ_i *are the field monomorphisms of the field* $\mathbb{Q}(\alpha)$. *Then*

$$|p + q_1\alpha + \cdots + q_m\alpha^m| > \frac{1}{(3mH)^m A^{m^2}}, \quad A = \max_{i=0,1,\ldots,m} \{1, |\alpha_i|\}, \tag{4.11}$$

for all

$$(p, q_1, \ldots, q_m)^t \in \mathbb{Z}^{m+1}, \quad 1 \le H = \max_{i=1,\ldots,m} |q_i|.$$

Proof We divide the proof in two cases. If $|p + q_1\alpha + \cdots + q_m\alpha^m| \ge 1$, we are done. From now on we suppose $|p + q_1\alpha + \cdots + q_m\alpha^m| < 1$. Immediately

$$|p| < 1 + |q_1\alpha + \cdots + q_m\alpha^m| \le 1 + mH \max\{1, |\alpha|\}^m$$

implying

$$\begin{aligned}
|p + q_1\alpha_i + \cdots + q_m\alpha_i^m| &< |p| + |q_1\alpha_i + \cdots + q_m\alpha_i^m| \\
&\le 1 + mH \max\{1, |\alpha|\}^m + mH \max\{1, |\alpha_i|\}^m \\
&\le 3mH \left(\max_{i=0,1,\ldots,m} \{1, |\alpha_i|\} \right)^m \\
&= 3mH A^m
\end{aligned}$$

Because $\alpha = \alpha_0$ is an algebraic integer of degree $\deg_{\mathbb{Q}} \alpha = m + 1$, then

$$\Theta := p + q_1\alpha + \cdots + q_m\alpha^m \in \mathbb{Z}[\alpha] \setminus \{0\}$$

is a non-zero algebraic integer and its field norm is an integer. Hence

$$1 \le |N(\Theta)| = |\Theta_0 \Theta_1 \cdots \Theta_m|$$

$$\le |\Theta| \prod_{i=1}^{m} |p + q_1 \alpha_i + \cdots + q_m \alpha_i^m|$$

$$\le |\Theta| (3mH)^m A^{m^2}. \quad \square$$

Theorem 4.7 *Let*

$$1, \alpha_1, \ldots, \alpha_m \in \mathbb{R}$$

be linearly independent over $\mathbb{Q}$. *If there exist positive constants* $c, \omega \in \mathbb{R}^+$ *such that*

$$|p_k + q_{1,k}\alpha_1 + \cdots + q_{m,k}\alpha_m| > \frac{c}{H_k^{\omega}}$$

for all

$$\bar{v}_k = (p_k, q_{1,k}, \ldots, q_{m,k})^t \in \mathbb{Z}^{m+1}, \quad 1 \le H_k = \max |q_{i,k}|,$$

then

$$\omega \ge m.$$

Finally, we note that by metrical considerations the upper bound in (4.10) may be improved for almost all m-tuples $(\alpha_1, \ldots, \alpha_m)$ of real numbers.

Theorem 4.8 *For almost all*

$$\alpha_1, \ldots, \alpha_m \in \mathbb{R}$$

wrt Lebesgue measure, there exist infinitely many primitive vectors

$$\bar{v}_k = (p_k, q_{1,k}, \ldots, q_{m,k})^t \in \mathbb{Z}^{m+1} \setminus \{\bar{0}\},$$

$$H_k := \max |q_{i,k}|, \ i = 1, \ldots, m,$$

satisfying

$$|p_k + q_{1,k}\alpha_1 + \cdots + q_{m,k}\alpha_m| < \frac{1}{H_k^m \log H_k}. \tag{4.12}$$

Further (4.12) is the best bound for a.a.

4.4 On Simultaneous Diophantine Inequalities

Theorem 4.9 is a variant of well-known simultaneous approximations.

Theorem 4.9 *Let*

$$\alpha_1, \ldots, \alpha_m \in \mathbb{R}, \quad f_1 + \cdots + f_m = 1,$$

be given. Then there exist

$$q \in \mathbb{Z}^+, \ p_1, \ldots, p_m \in \mathbb{Z}$$

satisfying

$$|q\alpha_i + p_i| < \frac{1}{q^{f_i}}, \quad \forall i = 1, \ldots, m. \tag{4.13}$$

Proof Define a set

$$\mathcal{C} = \{(x_0, x_1, \ldots, x_m)^t \in \mathbb{R}^{m+1} \mid |x_0| \leq q + 1/2, \ |x_0\alpha_i + x_i| < \frac{1}{q^{f_i}}, \ i = 1, \ldots, m\}.$$

The set $\mathcal{C}$ is a central symmetric convex body and its volume satisfies

$$V(\mathcal{C}) = (2q + 1)\frac{2}{q^{f_1}} \cdots \frac{2}{q^{f_m}} > 2^{m+1}.$$

The first Minkowski's convex body theorem with the lattice

$$\Lambda = \mathbb{Z}^{m+1}$$

gives a vector

$$\bar{0} \neq \bar{x} = (p, q_1, \ldots, q_m) \in \mathcal{C} \cap \Lambda.$$

Hence we have

$$(p, q_1, \ldots, q_m) \in \mathbb{Z}^{m+1} \setminus \{\bar{0}\}$$

satisfying the inequalities (4.13). $\square$

As immediate corollaries we get

Theorem 4.10 *Let*

$$\alpha_1, \ldots, \alpha_m \in \mathbb{R}$$

be given. Then there exist

$$q \in \mathbb{Z}^+, \ p_1, \ldots, p_m \in \mathbb{Z}$$

satisfying

$$|q\alpha_i + p_i| < \frac{1}{q^{1/m}}, \quad \forall i = 1, \ldots, m.$$

Theorem 4.11 *Let at least one of the numbers*

$$\alpha_1, \ldots, \alpha_m \in \mathbb{R}$$

be irrational. Then there exist infinitely many primitive vectors

$$\bar{v}_k = (q_k, p_{1,k}, \ldots, p_{m,k})^t \in \mathbb{Z}^{m+1} \setminus \{\bar{0}\}, \quad q_k \in \mathbb{Z}^+,$$

satisfying

$$|q_k \alpha_i + p_{i,k}| < \frac{1}{q_k^{1/m}}, \quad \forall i = 1, \ldots, m.$$

5 On Diophantine Inequalities Over $\mathbb{C}$

In the complex case Shidlovskii [12] studies linear forms over the ring of rational integers and gives the following

Theorem 5.1 ([12]) *Let* $\Theta_0 = 1, \Theta_1, \ldots, \Theta_m \in \mathbb{C}$ *and* $H \in \mathbb{Z}_{\geq 1}$ *be given. Then there exists a non-zero rational integer vector* $(\beta_0, \beta_1, \ldots, \beta_m)^t \in \mathbb{Z}^{m+1} \setminus \{\bar{0}\}$ *with* $|\beta_j| \leq H$, $j = 0, 1, \ldots, m$, *satisfying*

$$|\beta_0 + \beta_1 \Theta_1 + \cdots + \beta_m \Theta_m| \leq \frac{c}{H^{(m-1)/2}}, \quad c = \sqrt{2} \sum_{j=0}^{m} |\Theta_j|.$$

While Theorem 5.1 considers linear forms over rational integers only, the next result, see [8], is over the ring of integers $\mathbb{Z}_\mathbb{I}$ in an imaginary quadratic field $\mathbb{Q}(\sqrt{-D})$, $D \in \mathbb{Z}^+$, $D \not\equiv 0 \pmod 4$.

Theorem 5.2 ([8]) *Let* $\Theta_1, \ldots, \Theta_m \in \mathbb{C}$ *and* $H_1, \ldots, H_m \in \mathbb{Z}_{\geq 1}$ *be given. Then there exists a non-zero integer vector* $(\beta_0, \beta_1, \ldots, \beta_m)^t \in \mathbb{Z}_\mathbb{I}^{m+1} \setminus \{\bar{0}\}$, *with* $|\beta_j| \leq H_j$, $j = 1, \ldots, m$, *satisfying*

$$|\beta_0 + \beta_1 \Theta_1 + \cdots + \beta_m \Theta_m| \leq \left(\frac{2^\tau D^{1/4}}{\sqrt{\pi}} \right)^{m+1} \frac{1}{H_1 \cdots H_m}, \tag{5.1}$$

where $\tau = 1$, *if* $D \equiv 1$ *or* $2 \pmod 4$ *and* $\tau = 1/2$, *if* $D \equiv 3 \pmod 4$.

6 Siegel's Lemma

Siegel's lemma or Thue–Siegel's lemma is a powerful tool in transcendental number theory, see e.g. [2, 12, 13]. We may say that most transcendence proofs are based on Siegel's lemma.

Let $a_{mn} \in \mathbb{Z}$ ($m = 1, \ldots, M; n = 1, \ldots, N$) be given. Suppose $M < N$, then Siegel's lemma gives an estimate for the size of an integer solution $(x_1, \ldots, x_N)$ to the system of equations

$$\begin{cases} a_{11}x_1 + a_{12}x_2 + \cdots + a_{1N}x_N = 0, \\ a_{21}x_1 + a_{22}x_2 + \cdots + a_{2N}x_N = 0, \\ \cdots \\ a_{M1}x_1 + a_{M2}x_2 + \cdots + a_{MN}x_N = 0. \end{cases} \qquad (6.1)$$

Theorem 6.1 *Let*

$$L_m(\overline{x}) = \sum_{n=1}^{N} a_{mn}x_n, \quad m = 1, \ldots, M,$$

be M non-trivial linear forms with coefficients $a_{mn} \in \mathbb{Z}$ in N variables x_k. We also assume that

$$A_m := \sum_{n=1}^{N} |a_{mn}| \in \mathbb{Z}^+, \quad m = 1, \ldots, M. \qquad (6.2)$$

Suppose that

$$M < N,$$

then the system of equations

$$L_m(\overline{x}) = 0, \quad m = 1, \ldots, M, \qquad (6.3)$$

has a non-zero integer solution $\overline{z} = (z_1, \ldots, z_N)^t \in \mathbb{Z}^N \setminus \{\overline{0}\}$ with

$$1 \le \max_{1 \le n \le N} |z_n| \le \left\lfloor (A_1 \cdots A_M)^{\frac{1}{N-M}} \right\rfloor. \qquad (6.4)$$

Here and in the sequel $\lfloor x \rfloor$ denotes the largest integer $\le x$.

Note that the upper bound in (6.4) may further be estimated to give a solution with

$$1 \le \max_{1 \le n \le N} |z_n| \le \left(N \max_{1 \le m,n \le N} |a_{m,n}| \right)^{\frac{M}{N-M}}. \qquad (6.5)$$

6.1   Proof of Siegel's Lemma

Let

$$\mathcal{A} := [a_{mn}] \in M_{M \times N}(\mathbb{Z})$$

be the matrix of the $\mathbb{Z}$-module mapping

$$\overline{L} = (L_1, \ldots, L_M)^t : \mathbb{Z}^N \to \mathbb{Z}^M, \quad N > M. \qquad (6.6)$$

The homomorphism (6.6) between additive groups $(\mathbb{Z}^N, +)$ and $(\mathbb{Z}^M, +)$ is not injective and thus

$$\mathrm{Ker}\, \overline{L} \neq \{\overline{0}\}.$$

Thus there exists a non-zero integer solution $\overline{z} = (z_1, \ldots, z_N)^t \in \mathbb{Z}^N \setminus \{\overline{0}\}$ to (6.3). But we do not know its size.

In the following we shall prove the estimate given in (6.4). Denote

$$Z := \left\lfloor (A_1 \cdots A_M)^{\frac{1}{N-M}} \right\rfloor.$$

Then

$$A_1 \cdots A_M < (Z+1)^{N-M}, \tag{6.7}$$

where $Z, A_1, \ldots, A_M$ are positive integers. Consequently,

$$(A_1 Z + 1) \cdots (A_M Z + 1) \leq A_1 \cdots A_M (Z+1)^M < (Z+1)^N. \tag{6.8}$$

First we define a box

$$\square_1 := \{\overline{x} \in \mathbb{Z}^N \mid 0 \leq x_n \leq Z\},$$

where the number of integer points is

$$\#\square_1 = (Z+1)^N.$$

The linear mappings

$$L_m(\overline{x}) = \sum_{n=1}^{N} a_{mn} x_n, \quad m = 1, \ldots, M,$$

are bounded in the box $\square_1$ by

$$\sum_{a_{mn} < 0} a_{mn} x_n \leq L_m(\overline{x}) \leq \sum_{a_{mn} > 0} a_{mn} x_n.$$

Define

$$-b_m := \sum_{a_{mn} < 0} a_{mn}, \quad c_m := \sum_{a_{mn} > 0} a_{mn} x_n$$

and note

$$b_m + c_m = A_m.$$

Thus we have the estimates

$$-b_m Z \le L_m(\overline{x}) \le c_m Z.$$

Define the second box by

$$\square_2 := \{\overline{l} \in \mathbb{Z}^M | -b_m Z \le l_m \le c_m Z\},$$

where

$$\#\{l_m\} = (b_m + c_m)Z + 1 = A_m Z + 1.$$

The number of points in the second box $\square_2$ is

$$\#\square_2 = (A_1 Z + 1) \cdots (A_M Z + 1).$$

We have

$$\overline{L}(\square_1) \subseteq \square_2,$$

where

$$\#\square_2 = (A_1 Z + 1) \cdots (A_M Z + 1) < \#\square_1 = (Z + 1)^N$$

by (6.8). Hence

$$\overline{L} : \square_1 \to \square_2$$

is not injective on $\square_1$. Therefore there exist two different vectors $\overline{x}_1, \overline{x}_2 \in \square_1$ such that $\overline{L}(\overline{x}_1) = \overline{L}(\overline{x}_2)$, which further gives

$$\overline{L}(\overline{x}_1 - \overline{x}_2) = \overline{0}, \quad \overline{x}_1 - \overline{x}_2 \in \pm\square_1 \setminus \{\overline{0}\}.$$

By denoting

$$\overline{z} = (z_1, \ldots, z_N)^t := \overline{x}_1 - \overline{x}_2$$

we get a non-zero solution to (6.3) satisfying the estimates

$$-Z \le z_n \le Z, \quad Z = \left\lfloor (A_1 \cdots A_M)^{\frac{1}{N-M}} \right\rfloor, \quad n = 1, \ldots, N. \quad \square$$

6.2   Siegel's Lemma in Imaginary Quadratic Fields

There are several variations of Siegel's lemma over algebraic number fields, see e.g. [4]. Here we will mention a recent version, proved in [6], over imaginary quadratic fields. Let $\mathbb{I}$ denote the field $\mathbb{Q}$ of rational numbers or an imaginary quadratic field $\mathbb{Q}(\sqrt{-D})$, where $D \in \mathbb{Z}^+$ and $\mathbb{Z}_{\mathbb{I}}$ its ring of integers.

Theorem 6.2 ([6]) *Let*

$$L_m(\bar{z}) = \sum_{n=1}^{N} a_{mn} z_n, \quad m = 1, \ldots, M,$$

be M non-trivial linear forms with coefficients $a_{mn} \in \mathbb{Z}_\mathbb{I}$ in N variables z_n. Define $A_m := \sum_{n=1}^{N} |a_{mn}| \in \mathbb{Z}^+$ for $m = 1, \ldots, M$. Suppose that $M < N$. Then there exist positive constants $s_\mathbb{I}, t_\mathbb{I}$ such that the system of equations

$$L_m(\bar{z}) = 0, \quad m = 1, \ldots, M, \tag{6.9}$$

has a solution $\bar{z} = (z_1, \ldots, z_N)^t \in {\mathbb{Z}_\mathbb{I}}^N \setminus \{\bar{0}\}$ with

$$1 \leq \max_{1 \leq n \leq N} |z_n| \leq s_\mathbb{I} t_\mathbb{I}^{\frac{M}{N-M}} (A_1 \cdots A_M)^{\frac{1}{N-M}},$$

where $s_\mathbb{Q} = t_\mathbb{Q} = 1$ (see e.g. [7]) and $s_\mathbb{I}, t_\mathbb{I}$ are suitable constants depending on the field.

More precisely:

Lemma 6.3 ([6]) *There exists a solution to the system (6.9) of equations with*

$$1 \leq \max_{1 \leq n \leq N} |z_n| \leq \max \left\{ 2c\sqrt{D}, s_\mathbb{I} t_\mathbb{I}^{\frac{M}{N-M}} (A_1 \cdots A_M)^{\frac{1}{N-M}} \right\},$$

where

$$s_{\mathbb{Q}(\sqrt{-D})} = \begin{cases} \frac{2\sqrt{2}D^{1/4}}{\sqrt{\pi}}; \\ \frac{2}{\sqrt{\pi}}D^{1/4}; \end{cases} \quad t_{\mathbb{Q}(\sqrt{-D})} = \begin{cases} \frac{5}{2\sqrt{2}}, & D \equiv 1 \text{ or } 2 \pmod 4; \\ \frac{5}{2\sqrt{2}}, & D \equiv 3 \pmod 4; \end{cases}$$

and

$$c = \begin{cases} 2\sqrt{2}, & D \equiv 1 \text{ or } 2 \pmod 4; \\ 2, & D \equiv 3 \pmod 4. \end{cases}$$

6.2.1 An Application of Theorem 6.2

Theorem 6.2 is applied in [6] for studying Baker type lower bounds of linear forms of exponential function values (see also Baker [1]). From [6] we mention a corollary which gives a new generalised transcendence measure for e.

Theorem 6.4 *Let $m \in \mathbb{Z}_{\geq 2}$. Then*

$$|\beta_0 + \beta_1 e + \beta_1 e^2 + \cdots + \beta_m e^m| > \frac{1}{h^{1+\epsilon(h)}}, \quad h = h_1 \cdots h_m,$$

is valid for all $\overline{\beta} = (\beta_0, \ldots, \beta_m)^t \in \mathbb{Z}_{\mathbb{I}}^m \setminus \{\overline{0}\}$, $h_i = \max\{1, |\beta_i|\}$ *with*

$$\epsilon(h) = \frac{(4 + 7m)\sqrt{\log(m + 1)}}{\sqrt{\log \log h}},$$

$$\log h \geq m^2(41 \log(m + 1) + 10)e^{m^2(81 \log(m+1)+20)}.$$

7   Towards Height Theorem of Rational Subspaces

In the second part of our lectures our aim is to give a proof to the equality of heights of a rational subspace and its orthogonal complement, compare Schmidt [11]. The proof is based on Grassmann algebra. Therefore, quite an amount of basics of exterior algebras will be presented in the framework of rational subspaces. An important tool will be the primitive Grassmann coordinate vector of the corresponding rational subspace.

8   Towards Exterior Algebras

For a full understanding of exterior (wedge) products one needs some formal understanding of tensor products of modules, see Rotman [9], Sect. 8.4. However, in the following you may find a collection of necessary definitions and results which are quite enough for our purpose.

8.1   Algebras

Let R be a commutative ring. Four-tuple

$$(M, \times, +, \cdot)$$

is an *R-algebra*, if

$$(M, \times, +)$$

is a ring, $1 \in M$, $1 \times m = m$ for all $m \in M$, and

$$(M, +, \cdot)$$

is an R-module and moreover

$$a \cdot (m \times n) = (a \cdot m) \times n = m \times (a \cdot n), \ \forall a \in R, \ m, n \in M.$$

8.1.1 Exterior Algebra, Wedge Product

Definition 8.1 Let R be a commutative ring. An R-algebra

$$(E, \wedge, +, \cdot)$$

is an exterior algebra, if the ring-product $\wedge$, the wedge-product, is alternating:

$$m \wedge m = 0, \quad \forall m \in E. \tag{8.1}$$

The following lemma characterizes the alternating property.

Lemma 8.2 *Let M be an R-algebra. Then*

$$m \wedge m = 0 \ \forall m \in M \quad \Rightarrow \quad m \wedge n = -n \wedge m \ \ \forall m, n \in M. \tag{8.2}$$

If $char R \neq 2$, then

$$m \wedge n = -n \wedge m \ \ \forall m, n \in M \quad \Rightarrow \quad m \wedge m = 0 \ \forall m \in M. \tag{8.3}$$

Proof of (8.2) Let $m, n \in M$, then by the assumption in (8.2) we get

$$\begin{aligned}
0 &= (m + n) \wedge (m + n) \\
&= m \wedge m + m \wedge n + n \wedge m + n \wedge n \\
&= m \wedge n + n \wedge m.
\end{aligned}$$

Hence $m \wedge n = -n \wedge m$ follows. $\square$
The proof of (8.3) is left as an exercise.

8.2 Tensor Product

Definition 8.3 A tensor product of R-modules M and N is a pair formed by a group

$$M \otimes_R N$$

with an R-bilinear function

$$h : M \times N \to M \otimes_R N$$

satisfying: for every R-module G and every R-bilinear function

$$f : M \times N \to G$$

there exists a unique R-bilinear function such that

$$\tilde{f} : M \otimes_R N \to G, \quad f = \tilde{f} \circ h. \tag{8.4}$$

As usual, the Definition 8.3 may shortly be given as a commutative diagram

$$\begin{array}{ccc}
 & R\text{-bilinear } h & \\
M \times N & \to & M \otimes_R N \\
\forall f \searrow & & \swarrow \exists! \tilde{f} : f = \tilde{f} \circ h \\
 & \forall G &
\end{array}$$

Let R be a commutative ring. The elements of the tensor product $M \otimes_R N$ are denoted by

$$m \otimes n \in M \otimes_R N, \quad m \in M, n \in N,$$

and they satisfy the following rules

$$m \otimes (n_1 + n_2) = m \otimes n_1 + m \otimes n_2, \quad (m_1 + m_2) \otimes n = m_1 \otimes n + m_2 \otimes n,$$

$$r \cdot m \otimes n = (r \cdot m) \otimes n = m \otimes (r \cdot n), \quad \forall r \in R, m \in M, n \in N.$$

The above definition is quite general but it is not very illustrative in the beginning. In the following we give a couple of examples which illustrate the idea of tensor product very well.

8.2.1 Tensor Product of Free Modules M and N Over the Commutative Ring R

If $\text{rank}_R M = m$, $\text{rank}_R N = n$ and

$$M = Ra_1 + \cdots + Ra_m, \quad N = Rb_1 + \cdots + Rb_n,$$

then

$$\text{rank}_R M \otimes_R N = mn$$

and

$$M \otimes_R N = Ra_1 \otimes b_1 + Ra_1 \otimes b_2 + \cdots + Ra_m \otimes b_n.$$

So, the tensor product $M \otimes_R N$ of modules is spanned by the tensor products of all pairs of base vectors from M and N.

8.2.2 Tensor Product of Vector Spaces M and N Over the Field K

If $\dim_K M = m$, $\dim_K N = n$ and

$$M = Ka_1 + \cdots + Ka_m, \quad N = Kb_1 + \cdots + Kb_n,$$

then

$$\dim_K M \otimes_K N = mn$$

and

$$M \otimes_K N = Ka_1 \otimes b_1 + Ka_1 \otimes b_2 + \cdots + Ka_m \otimes b_n.$$

Thus, the tensor product gives a formal way to multiply vector spaces.

8.2.3 Extending Scalars

Recall that an R-module M is free, if $M \cong R^m$.

Let M be a free R-module and R be a subring of the ring S. Then

$$S \otimes_R M$$

gives a module extending the scalars from R to S. Namely, if $\operatorname{rank}_R M = m$, then

$$M \cong R^m, \quad S \otimes_R M \cong S^m.$$

8.2.4 Tensor Algebra

Let M be an R-module. The tensor algebra $T(M)$ on M is defined by setting

$$T(M) = \sum_{p \geq 0} T^p(M),$$

where

$$T^0(M) = R,$$
$$T^1(M) = M,$$
$$T^p(M) = M \otimes \cdots \otimes M = \langle m_1 \otimes \cdots \otimes m_p \mid m_1, \ldots, m_p \in M \rangle, \quad p \geq 2.$$

8.2.5 Construction of Exterior Algebra

Now we are ready to construct exterior algebra on an R-module M by using the above tensor algebra on M. We get the exterior algebra on M by a quotient space

$$\bigwedge(M) := T(M)/J,$$

where J is an ideal generated by all $m \otimes m$ with $m \in M$. Let

$$m_1 \otimes \cdots \otimes m_p \to m_1 \wedge \cdots \wedge m_p$$

be the corresponding canonical map. Then the product $\wedge$ is alternating.

Example 8.4 Let M be an R-module, then

$$\begin{cases} m_1 \wedge m_2 \wedge m_2 = 0, \\ m_3 \wedge m_2 \wedge m_1 = -m_1 \wedge m_2 \wedge m_3, \end{cases}$$

for all $m_1, m_2, m_3 \in M$.

9 Grassmann Algebra

9.1 Basics

Assume that the R-module M has base

$$\bar{e}_1, \ldots, \bar{e}_n \in M, \quad \text{rank}_R M = n,$$

which allows us to write

$$M = \langle \bar{e}_1, \ldots, \bar{e}_n \rangle = R\bar{e}_1 + \cdots + R\bar{e}_n.$$

In the exterior algebra $\bigwedge(M)$ we have following properties.

Lemma 9.1 *Let τ be a permutation in a set $\{1, 2, \ldots, n\}$, then*

$$\bar{e}_{\tau(1)} \wedge \cdots \wedge \bar{e}_{\tau(n)} = sign(\tau)\bar{e}_1 \wedge \cdots \wedge \bar{e}_n, \tag{9.1}$$

where $sign(\tau)$ is the signature of τ. Moreover

$$\bar{e}_{i_1} \wedge \cdots \wedge \bar{e}_{i_p} \neq \bar{0}, \quad \forall\, 1 \leq i_1 < i_2 < \cdots < i_p \leq n, \tag{9.2}$$

and

$$r \cdot \overline{X} = \overline{0}, \quad r \in R, \ \overline{X} \in \bigwedge(M) \quad \Rightarrow \quad r = 0 \ \text{ or } \ \overline{X} = \overline{0}. \qquad (9.3)$$

9.2 Exterior Base

For studying exterior products we introduce the concept of an increasing $0 \le p \le n$-list.

Definition 9.2 Let $n \in \mathbb{Z}_{\ge 1}$. An increasing $0 \le p \le n$-list σ_p is a list

$$\begin{cases} \sigma = \sigma_p := i_1, \ldots, i_p; \quad 1 \le i_1 < i_2 < \cdots < i_p \le n, \quad \text{if} \quad p \ge 1; \\ \sigma := \emptyset, \quad \text{if} \quad p = 0. \end{cases}$$

Further, corresponding to a $0 \le p \le n$-list σ_p, we introduce a p-wedge-product

$$\overline{E}_\sigma := \overline{e}_{i_1} \wedge \cdots \wedge \overline{e}_{i_p}, \quad p \ge 1,$$

$$\overline{E}_\emptyset := 1, \quad p = 0.$$

The set of all $0 \le p \le n$-lists is denoted by $C(n, p) := \{\sigma_p\}$. Obviously, $\#C(n, p) = \binom{n}{p}$.

Now we suppose that R is an integral domain and put $M = R^n$ (a free R-module). If $R = K$ is a field, then $M = R^n$ is a vector space. Let $0 \le p \le n$. Then we may define a free R-module R^n_p by the linear hull

$$R^n_p = \langle \overline{E}_{\sigma_p} \mid \sigma_p \in C(n, k) \rangle_R, \quad \text{rank}_R R^n_p = \binom{n}{p}$$

generated by all p-wedge-products $\overline{E}_{\sigma_p} = \overline{e}_{i_1} \wedge \cdots \wedge \overline{e}_{i_p}$. In particular

$$R^n_0 = \langle \overline{E}_\emptyset \rangle_R, \quad \text{rank}_R R^n_0 = 1;$$

$$R^n_1 = \langle \overline{E}_\sigma \in \{\overline{e}_1, \ldots, \overline{e}_n\} \rangle_R, \quad \text{rank}_R R^n_1 = n;$$

$$R^n_2 = \langle \overline{E}_\sigma \in \{\overline{e}_1 \wedge \overline{e}_2, \ldots\} \rangle_R, \quad \text{rank}_R R^n_2 = \binom{n}{2}.$$

$$R^n_n = \langle \overline{E}_\sigma = \overline{e}_1 \wedge \overline{e}_2 \wedge \cdots \wedge \overline{e}_{n-1} \wedge \overline{e}_n \rangle_R, \quad \text{rank}_R R^n_n = 1.$$

Now we are ready to define Grassmann algebra.

Definition 9.3 The Grassmann algebra is a graded R-algebra

$$G_n := R_0^n \oplus \cdots \oplus R_n^n, \quad \mathrm{rank}_R G_n = 2^n.$$

Note that Grassmann algebra is an exterior algebra over R, hence an R-module, too.

10 Determinant

Let M be an R-module of rank $r(M) = n$. A mapping

$$f : M \to M, \quad f(a \cdot m) = a \cdot f(M), \quad f(m+n) = f(n) + f(m)$$

is called an R-homomorphism. By using the functor

$$\bigwedge^n (f) : \overline{x}_1 \wedge \cdots \wedge \overline{x}_n \to f(\overline{x}_1) \wedge \cdots \wedge f(\overline{x}_n),$$

we may define the determinant of f.

Definition 10.1 The determinant of f is an element $\det f \in R$ satisfying

$$\bigwedge^n (f) \overline{e}_1 \wedge \cdots \wedge \overline{e}_n = f(\overline{e}_1) \wedge \cdots \wedge f(\overline{e}_n) := \det(f) \cdot \overline{e}_1 \wedge \cdots \wedge \overline{e}_n. \quad (10.1)$$

Example 10.2 Let $n = p = 2$ and let f be the R-homomorphism defined by

$$f(\overline{e}_1) = a_{11} \overline{e}_1 + a_{21} \overline{e}_2, \quad f(\overline{e}_2) = a_{12} \overline{e}_1 + a_{22} \overline{e}_2.$$

Then, by exterior algebra

$$(a_{11} \overline{e}_1 + a_{21} \overline{e}_2) \wedge (a_{12} \overline{e}_1 + a_{22} \overline{e}_2) = (a_{11} a_{22} - a_{12} a_{21}) \cdot \overline{e}_1 \wedge \overline{e}_2$$

which implies

$$\det(f) = a_{11} a_{22} - a_{12} a_{21}.$$

By denoting

$$\overline{a}_1 := A \overline{e}_1 = a_{11} \overline{e}_1 + a_{21} \overline{e}_2, \quad \overline{a}_2 := A \overline{e}_2 = a_{12} \overline{e}_1 + a_{22} \overline{e}_2,$$

we get a matrix representation

$$A = [\overline{a}_1, \overline{a}_2], \quad \det A := \det f = a_{11} a_{22} - a_{12} a_{21},$$

for the above R-homomorphism f and its determinant. More generally, we denote

$$\bar{a}_i := A\bar{e}_i = a_{1i}\bar{e}_1 + \cdots + a_{ni}\bar{e}_n, \quad i = 1, \ldots, n,$$

and hence

$$A = [\bar{a}_1, \ldots, \bar{a}_n].$$

10.1 Expansions

Let

$$H = H_p = h_1, \ldots, h_p, \quad 1 \le h_1 < \cdots < h_p \le n,$$

$$K = K_q = k_1, \ldots, k_q, \quad 1 \le k_1 < \cdots < k_q \le n,$$

be increasing $0 \le p \le n$ and $0 \le q \le n$-lists, respectively. If the lists are disjoint and $0 \le p+q \le n$, let $\tau_{H,K}$ be a permutation which arranges $h_1, \ldots, h_p, k_1, \ldots, k_q$ into an increasing $0 \le p+q \le n$-list

$$H * K := j_1, \ldots, j_{p+q}, \quad 1 \le j_1 < \cdots < j_{p+q} \le n.$$

Lemma 10.3 *Then*

$$\bar{E}_H \wedge \bar{E}_K = \begin{cases} 0, & \text{if } H \cap K \ne \emptyset; \\ sign(\tau_{H,K})\bar{E}_{H*K}, & \text{if } H \cap K = \emptyset. \end{cases} \tag{10.2}$$

In the following we use also the notation $H = h_1 < \cdots < h_p$ for an increasing list $H = h_1, \ldots, h_p, \quad 1 \le h_1 < \cdots < h_p \le n$.

Example 10.4 By (10.2) we get

$$(\bar{e}_1 \wedge \bar{e}_5) \wedge (\bar{e}_2 \wedge \bar{e}_4 \wedge \bar{e}_6) = +\bar{e}_1 \wedge \bar{e}_2 \wedge \bar{e}_4 \wedge \bar{e}_5 \wedge \bar{e}_6,$$

where

$$H = H_2 = 1 < 5, \quad K = K_3 = 2 < 4 < 6,$$

$$H * K = 1 < 2 < 4 < 5 < 6, \quad n = 6, \quad p + q = 5.$$

Now we consider certain submatrices of

$$A = [a_{st}] = [\bar{a}_1, \ldots, \bar{a}_n] \in M_{n \times n}.$$

Corresponding to the increasing lists $H = h_1, \ldots, h_p$, $K = k_1, \ldots, k_p \in C(n, p)$ we define a $p \times p$ submatrix

$$A_{HK} := [a_{st}], \quad (s, t) \in H \times K.$$

The determinant $\det(A_{HK})$ is called a minor of A of order p.

Lemma 10.5 (Lemma 9.158 in [9]) *Corresponding to the increasing list $H = h_1 < \cdots < h_p$, there is an expansion*

$$\bar{a}_{h_1} \wedge \cdots \wedge \bar{a}_{h_p} = \sum_{L \in C(n,p)} \det(A_{L,H}) \bar{E}_L, \tag{10.3}$$

where the scalars $\det(A_{L,H})$ are so called Grassmann or Plücker coordinates of $\bar{a}_{h_1} \wedge \cdots \wedge \bar{a}_{h_p}$.

Proof Write $L = l_1 < \cdots < l_p$, then we have

$$\bar{a}_{h_1} \wedge \cdots \wedge \bar{a}_{h_p} = (A\bar{e}_{h_1}) \wedge \cdots \wedge (A\bar{e}_{h_p})$$

$$= \left(\sum_{t_1=1,\ldots,n} a_{t_1,h_1} \bar{e}_{t_1} \right) \wedge \cdots \wedge \left(\sum_{t_p=1,\ldots,n} a_{t_p,h_p} \bar{e}_{t_p} \right)$$

$$= \sum_{T=t_1,\ldots,t_p} a_{t_1,h_1} \cdots a_{t_p,h_p} \cdot \bar{e}_{t_1} \wedge \cdots \wedge \bar{e}_{t_1}$$

$$= \sum_{L=l_1<\cdots<l_p \in C(n,p)} \left(\sum_{\tau:T\to L} \mathrm{sign}(\tau) a_{\tau^{-1}(l_1),h_1} \cdots a_{\tau^{-1}(l_p),h_p} \right) \cdot \bar{e}_{l_1} \wedge \cdots \wedge \bar{e}_l$$

$$= \sum_{L \in C(n,p)} \det(A_{L,H}) \cdot \bar{E}_L. \quad \square$$

Example 10.6 Let $p = 2, n = 3$ and consider the wedge product $\bar{a}_1 \wedge \bar{a}_2$ of

$$\bar{a}_1 = a_{11}\bar{e}_1 + a_{21}\bar{e}_2 + a_{31}\bar{e}_3,$$

$$\bar{a}_2 = a_{12}\bar{e}_1 + a_{22}\bar{e}_2 + a_{32}\bar{e}_3.$$

After some rearrangement we may write

$$\bar{a}_1 \wedge \bar{a}_2 = \begin{vmatrix} a_{11} & a_{12} \\ a_{21} & a_{22} \end{vmatrix} \bar{e}_1 \wedge \bar{e}_2 + \begin{vmatrix} a_{11} & a_{12} \\ a_{31} & a_{32} \end{vmatrix} \bar{e}_1 \wedge \bar{e}_3 + \begin{vmatrix} a_{21} & a_{22} \\ a_{31} & a_{32} \end{vmatrix} \bar{e}_2 \wedge \bar{e}_3$$

$$= \det(A_{L_1,H}) \bar{E}_{L_1} + \det(A_{L_2,H}) \bar{E}_{L_2} + \det(A_{L_3,H}) \bar{E}_{L_3},$$

where

$$L_1 = 1 < 2, \quad L_2 = 1 < 3, \quad L_3 = 2 < 3, \quad H = 1 < 2.$$

Hereafter, corresponding to the increasing lists $I = i_1 < \cdots < i_p$, we write

$$\overline{A}_I = \overline{a}_{i_1} \wedge \cdots \wedge \overline{a}_{i_p}.$$

Lemma 10.7 (9.160(i) in [9]) *Let $I = i_1, \ldots, i_p$ and $J = j_1, \ldots, j_q$ be increasing lists, where $p, q \in \{1, \ldots, n\}$. Then*

$$\overline{A}_I \wedge \overline{A}_J = \sum_{H \in C(n,p), K \in C(n,q)} sign(\tau_{H,K}) \det(A_{H,I}) \det(A_{K,J}) \overline{E}_{H*K}.$$

Proof

$$
\begin{aligned}
\overline{A}_I \wedge \overline{A}_J &= \sum_H \det(A_{H,I}) \overline{E}_H \wedge \sum_K \det(A_{K,J}) \overline{E}_K \\
&= \sum_{H,K} \det(A_{H,I}) \det(A_{K,J}) \overline{E}_H \wedge \overline{E}_K \\
&= \sum_{H,K} sign(\tau_{H,K}) \det(A_{H,I}) \det(A_{K,J}) \overline{E}_{H*K}. \quad \square
\end{aligned}
$$

If $I * J = 1 < 2 < \cdots < n = p + q$, then the above gives

$$\overline{a}_{i_1} \wedge \cdots \wedge \overline{a}_{i_p} \wedge \overline{a}_{j_1} \wedge \cdots \wedge \overline{a}_{j_q}$$

$$= \sum_{H \in C(n,p), K \in C(n,q)} sign(\tau_{H,K}) \det(A_{H,I}) \det(A_{K,J}) \overline{E}_{H*K}.$$

Thus immediate corollaries are

$$\det A = sign(\tau_{I,J}) \sum_{H,K} sign(\tau_{H,K}) \det(A_{H,I}) \det(A_{K,J})$$

and

$$\det[a_{hi}]_{n \times n} = \sum_{h=1,\ldots,n} (-1)^{h+i} a_{hi} \det(A_{\hat{h},\hat{i}}), \quad i = 1, \ldots, n,$$

where $\hat{h} = 1 < 2 < a < h - 1 < h + 1 < \cdots \le n$.

10.2 Linear Independence

In this chapter we assume that R is a field and M is an n-dimensional vector space over R. Note that most of the following results are still valid in R-modules, where R

is a commutative ring with unity. The vectors of M are denoted by $\bar{x}_i, \bar{y}_j, \ldots \in M$. The notation $S^\perp$ is used for the orthogonal complement of the subspace $S \subseteq M$.

Lemma 10.8 (Lemma 5 C in [11]) *Let $\bar{x}_1, \ldots, \bar{x}_p \in M$. Then*

$$\bar{x}_1, \ldots, \bar{x}_p \quad \text{are linearly dependent over } R \quad \Leftrightarrow \quad \bar{x}_1 \wedge \cdots \wedge \bar{x}_p = \bar{0}. \quad (10.4)$$

Further,

$$\bar{z} \in \langle \bar{x}_1, \ldots, \bar{x}_p \rangle \quad \Leftrightarrow \quad \bar{z} \wedge \bar{x}_1 \wedge \cdots \wedge \bar{x}_p = \bar{0}. \quad (10.5)$$

We shall use the notations

$$\underset{p}{\wedge}\overline{X} = \bar{x}_1 \wedge \cdots \wedge \bar{x}_p;$$

$$\hat{\wedge}_k \overline{X} = \bar{x}_1 \wedge \cdots \wedge \bar{x}_{k-1} \wedge \bar{x}_{k+1} \wedge \cdots \wedge \bar{x}_p \cdot (-1)^{p-k}.$$

Readily

$$\hat{\wedge}_k \overline{X} \wedge \bar{x}_j = \delta_{kj} \cdot \underset{p}{\wedge}\overline{X}.$$

Proof of Lemma 10.8 Suppose that $\bar{x}_1, \ldots, \bar{x}_p$ are linearly dependent over R. Then there exist a linear combination

$$r_1 \cdot \bar{x}_1 + \cdots + r_k \cdot \bar{x}_k + \cdots + r_p \cdot \bar{x}_p = \bar{0}, \quad r_1, \ldots, r_p \in R, \quad r_k \neq 0. \quad (10.6)$$

Operate (10.6) by $\hat{\wedge}_k \overline{X} \wedge$, then

$$r_1 \cdot \bar{0} + \cdots + r_k \cdot \underset{p}{\wedge}\overline{X} + \cdots + r_p \cdot \bar{0} = \bar{0}$$

implying

$$\underset{p}{\wedge}\overline{X} = \bar{x}_1 \wedge \cdots \wedge \bar{x}_p = \bar{0}.$$

Suppose then

$$\bar{x}_1 \wedge \cdots \wedge \bar{x}_p = \bar{0}. \quad (10.7)$$

Assume on the contrary that

$$\dim_R S = p, \quad S := \langle \bar{x}_1, \ldots, \bar{x}_p \rangle_R.$$

Then

$$\dim_R S^\perp = n - p, \quad S^\perp = \langle \bar{z}_1, \ldots, \bar{z}_{n-p} \rangle,$$

with some $\bar{z}_1, \ldots, \bar{z}_{n-p} \in M$. Thus

$$\dim_R \langle \bar{x}_1, \ldots, \bar{x}_p, \bar{z}_1, \ldots, \bar{z}_{n-p} \rangle_R = n.$$

Now by (10.7) we get

$$\bar{0} = \bar{x}_1 \wedge \cdots \wedge \bar{x}_p \wedge \bar{z}_1 \wedge \cdots \wedge \bar{z}_{n-p} = \det[\bar{x}_1, \ldots, \bar{x}_p, \bar{z}_1, \ldots, \bar{z}_{n-p}] \cdot \bar{e}_1 \wedge \cdots \wedge \bar{e}_n,$$

implying

$$\det[\bar{x}_1, \ldots, \bar{x}_p, \bar{z}_1, \ldots, \bar{z}_{n-p}] = 0.$$

Then by linear algebra

$$\dim_R \langle \bar{x}_1, \ldots, \bar{x}_p, \bar{z}_1, \ldots, \bar{z}_{n-p} \rangle_R \leq n - 1.$$

A contradiction. This proves (10.4).

The case (10.5) is left as an exercise. $\square$

Lemma 10.9 (Lemma 5D in [11]) *Suppose $\bar{x}_1, \ldots, \bar{x}_p$ are linearly independent over R and $\bar{y}_1, \ldots, \bar{y}_p$ are linearly independent over R. Then*

$$\langle \bar{x}_1, \ldots, \bar{x}_p \rangle = \langle \bar{y}_1, \ldots, \bar{y}_p \rangle \quad \Leftrightarrow \quad \bar{x}_1 \wedge \cdots \wedge \bar{x}_p = \lambda \cdot \bar{y}_1 \wedge \cdots \wedge \bar{y}_p, \quad \lambda \in R.$$

Proof Suppose

$$\langle \bar{x}_1, \ldots, \bar{x}_p \rangle = \langle \bar{y}_1, \ldots, \bar{y}_p \rangle := N$$

where N is a p-dimensional R-vector space with a base $\bar{x}_1, \ldots, \bar{x}_p$. Now the vectors $\bar{y}_1, \ldots, \bar{y}_p$ may be written in the base $\bar{x}_1, \ldots, \bar{x}_p$ and thus by the definition of determinant

$$\bar{y}_1 \wedge \cdots \wedge \bar{y}_p = \det[\bar{y}_1, \ldots, \bar{y}_p]_{p \times p} \cdot \bar{x}_1 \wedge \cdots \wedge \bar{x}_p.$$

Assume

$$\bar{x}_1 \wedge \cdots \wedge \bar{x}_p = \lambda \cdot \bar{y}_1 \wedge \cdots \wedge \bar{y}_p, \quad \lambda \in R.$$

The wedge-product with $\bar{y}_j$ gives

$$\bar{y}_j \wedge \bar{x}_1 \wedge \cdots \wedge \bar{x}_p = \lambda \cdot \bar{y}_j \wedge \bar{y}_1 \wedge \cdots \wedge \bar{y}_p = \bar{0}.$$

Now Lemma 10.8 implies $\bar{y}_j \in \langle \bar{x}_1, \ldots, \bar{x}_p \rangle$ and so

$$\langle \bar{x}_1, \ldots, \bar{x}_p \rangle = \langle \bar{y}_1, \ldots, \bar{y}_p \rangle. \quad \square$$

11   Inner Products

In the following we consider vector spaces $M = R^n$ over the real $R = \mathbb{R}$ or complex field $R = \mathbb{C}$. The complex conjugates of scalars $r \in R$ and vectors $\overline{x} \in R^n$ are denoted by $\overline{r} \in R$ and $\overline{\overline{x}} \in R^n$, respectively.

The Hermitian inner product in R^n is defined in a usual manner by

$$\overline{x} \cdot \overline{y} = (x_1, \ldots, x_n)^t \cdot (y_1, \ldots, y_n)^t := x_1 \overline{y_1} + \cdots + x_n \overline{y_n}.$$

Further, the Hermitian inner product in R_p^n is analogously given by

$$\overline{Z} \odot \overline{W} = \sum_{H \in C(n,p)} Z_H \overline{E}_H \odot \sum_{K \in C(n,p)} W_K \overline{E}_K := \sum_{H \in C(n,p)} Z_H \overline{W_H}.$$

In particular,

$$\overline{E}_H \odot \overline{E}_K = \delta_{HK} = \begin{cases} 1, & H = K; \\ 0, & \text{otherwise}; \end{cases}$$

hold for the Hermitian inner product with the properties

$$(\overline{Y} + \overline{Z}) \odot \overline{W} = \overline{Y} \odot \overline{W} + \overline{Z} \odot \overline{W}; \quad (a \cdot \overline{Z}) \odot \overline{W} = a \cdot (\overline{Z} \odot \overline{W}).$$

Finally, the length (norm) $||\overline{Z}||$ of $\overline{Z} = \sum_{H \in C(n,p)} Z_H \overline{E}_H \in R_p^n$ is given by

$$||\overline{Z}||^2 := \overline{Z} \odot \overline{Z} = \sum_{H \in C(n,p)} |Z_H|^2.$$

11.1   Laplace Identities

Define a p-dimensional parallelepiped

$$F := [0, 1]\overline{x}_1 + \cdots + [0, 1]\overline{x}_p$$

spanned by $\overline{x}_1, \ldots, \overline{x}_p \in R^n$. Here we note that

$$V(F) = ||\overline{x}_1 \wedge \cdots \wedge \overline{x}_p||$$

gives the volume of F.

Lemma 11.1 (Laplace identities, Lemma 5E in [11]) *Let* $\overline{x}_1, \ldots, \overline{x}_p, \overline{y}_1, \ldots, \overline{y}_p \in R^n$. *Then*

$$(\bar{x}_1 \wedge \ldots \wedge \bar{x}_p) \odot (\bar{y}_1 \wedge \ldots \wedge \bar{y}_p) = \begin{vmatrix} \bar{x}_1 \cdot \bar{y}_1 & \ldots & \bar{x}_1 \cdot \bar{y}_p \\ \cdot & & \cdot \\ \cdot & \cdot & \cdot \\ \cdot & & \cdot \\ \bar{x}_p \cdot \bar{y}_1 & \ldots & \bar{x}_p \cdot \bar{y}_p \end{vmatrix} \tag{11.1}$$

and

$$\|\bar{x}_1 \wedge \ldots \wedge \bar{x}_p\| = \begin{vmatrix} \bar{x}_1 \cdot \bar{x}_1 & \ldots & \bar{x}_1 \cdot \bar{x}_p \\ \cdot & & \cdot \\ \cdot & \cdot & \cdot \\ \cdot & & \cdot \\ \bar{x}_p \cdot \bar{x}_1 & \ldots & \bar{x}_p \cdot \bar{x}_p \end{vmatrix}^{1/2} . \tag{11.2}$$

Sketch of the proof Let $\bar{e}_1, \ldots, \bar{e}_n \in R^n$ be an orthonormal base of R^n. On the left-hand side of (11.1) we then have

$$\delta_{IJ} = \bar{E}_I \odot \bar{E}_J = (\bar{e}_{i_1} \wedge \cdots \wedge \bar{e}_{i_p}) \odot (\bar{e}_{j_1} \wedge \cdots \wedge \bar{e}_{j_p}),$$

and on the right-hand side of (11.1) we get

$$\begin{vmatrix} \bar{e}_{i_1} \cdot \bar{e}_{j_1} & \ldots & \bar{e}_{i_1} \cdot \bar{e}_{j_p} \\ \cdot & \cdot & \cdot \\ \cdot & \cdot & \cdot \\ \cdot & \cdot & \cdot \\ \bar{e}_{i_p} \cdot \bar{e}_{j_1} & \ldots & \bar{e}_{i_p} \cdot \bar{e}_{j_p} \end{vmatrix} = \delta_{IJ}.$$

Secondly, the left-hand and right-hand sides of (11.1) are multilinear. For example, on the left-hand side

$$((\bar{a}_1 + \bar{a}_2) \wedge \bar{x}_2 \wedge \cdots \wedge \bar{x}_p) \odot (\bar{y}_1 \wedge \cdots \wedge \bar{y}_p)$$
$$= (\bar{a}_1 \wedge \bar{x}_2 \wedge \cdots \wedge \bar{x}_p) \odot (\bar{y}_1 \wedge \cdots \wedge \bar{y}_p) + (\bar{a}_2 \wedge \bar{x}_2 \wedge \cdots \wedge \bar{x}_p) \odot (\bar{y}_1 \wedge \cdots \wedge \bar{y}_p).$$

And on the right-hand side

$$\begin{vmatrix} (\bar{a}_1 + \bar{a}_2) \cdot \bar{y}_1 & \ldots & (\bar{a}_1 + \bar{a}_2) \cdot \bar{y}_p \\ \cdot & & \cdot \\ \cdot & \cdot & \cdot \\ \cdot & & \cdot \\ \bar{x}_p \cdot \bar{y}_1 & \ldots & \bar{x}_p \cdot \bar{y}_p \end{vmatrix} = \begin{vmatrix} \bar{a}_1 \cdot \bar{y}_1 + \bar{a}_2 \cdot \bar{y}_1 & \ldots & \bar{a}_1 \cdot \bar{y}_p + \bar{a}_2 \cdot \bar{y}_p \\ \cdot & & \cdot \\ \cdot & \cdot & \cdot \\ \cdot & & \cdot \\ \bar{x}_p \cdot \bar{y}_1 & \ldots & \bar{x}_p \cdot \bar{y}_p \end{vmatrix}$$

$$= \begin{vmatrix} \bar{a}_1 \cdot \bar{y}_1 & \ldots & \bar{a}_1 \cdot \bar{y}_p \\ \cdot & & \cdot \\ \cdot & \cdot & \cdot \\ \cdot & & \cdot \\ \bar{x}_p \cdot \bar{y}_1 & \ldots & \bar{x}_p \cdot \bar{y}_p \end{vmatrix} + \begin{vmatrix} \bar{a}_2 \cdot \bar{y}_1 & \ldots & \bar{a}_2 \cdot \bar{y}_p \\ \cdot & & \cdot \\ \cdot & \cdot & \cdot \\ \cdot & & \cdot \\ \bar{x}_p \cdot \bar{y}_1 & \ldots & \bar{x}_p \cdot \bar{y}_p \end{vmatrix} . \qquad \square$$

11.2 Norm Inequality

Let $I = i_1, \ldots, i_p$ and $J = j_1, \ldots, j_q$ be increasing lists, where $p, q \in \{1, \ldots, n\}$ and write

$$\overline{A}_I = \overline{a}_{i_1} \wedge \cdots \wedge \overline{a}_{i_p} = \sum_{H \in C(n,p)} \det(A_{H,I}) \overline{E}_H,$$

$$\overline{A}_J = \overline{a}_{j_1} \wedge \cdots \wedge \overline{a}_{j_q} = \sum_{K \in C(n,q)} \det(A_{K,J}) \overline{E}_K.$$

Remember that Lemma 10.7 gives

$$\overline{A}_I \wedge \overline{A}_J = \overline{a}_{i_1} \wedge \cdots \wedge \overline{a}_{i_p} \wedge \overline{a}_{j_1} \wedge \cdots \wedge \overline{a}_{j_q}$$

$$= \sum_{H \in C(n,p), K \in C(n,q)} \text{sign}(\tau_{H,K}) \det(A_{H,I}) \det(A_{K,J}) \overline{E}_{H*K},$$

where

$$\overline{E}_H \wedge \overline{E}_K = \begin{cases} 0, & \text{if } H \cap K \neq \emptyset; \\ \text{sign}(\tau_{H,K}) \overline{E}_{H*K}, & \text{if } H \cap K = \emptyset. \end{cases} \tag{11.3}$$

The following lemma provides an upper bound for the length $||\overline{A}_I \wedge \overline{A}_J||$ in terms of the lengths $||\overline{A}_I||$ and $||\overline{A}_J||$.

Lemma 11.2 (Lemma 5F in [11]) *Let* $I = i_1, \ldots, i_p$ *and* $J = j_1, \ldots, j_q$ *be increasing lists, where* $p, q \in \{1, \ldots, n\}$. *Then*

$$||\overline{A}_I \wedge \overline{A}_J|| \leq ||\overline{A}_I|| ||\overline{A}_J|| \tag{11.4}$$

and if

$$\overline{a}_{i_c} \cdot \overline{a}_{j_d} = 0, \quad \forall\, i_c \in I, \; j_d \in J, \tag{11.5}$$

then equality holds in (11.4).

Note that

$$||\overline{A}_I||^2 = \sum_{H \in C(n,p)} |\det(A_{H,I})|^2, \quad ||\overline{A}_J||^2 = \sum_{K \in C(n,q)} |\det(A_{K,J})|^2.$$

11.3 Orthogonal Complement

Let $I = i_1, \ldots, i_p$ and $J = j_1, \ldots, j_q$ be increasing lists, where $p, q \in \{1, \ldots, n\}$ and let

$$S = \langle \bar{s}_{i_1}, \ldots, \bar{s}_{i_p} \rangle_R \subseteq R^n \tag{11.6}$$

be a p-dimensional subspace of R^n spanned by $\bar{s}_{i_1}, \ldots, \bar{s}_{i_p}$. Then the orthogonal complement

$$T := S^{\perp} = \{\bar{y} \in R^n \mid \bar{y} \cdot \bar{x} = 0, \ \forall \bar{x} \in S\}, \quad q := n - p, \tag{11.7}$$

is a q-dimensional subspace

$$T = \langle \bar{t}_{j_1}, \ldots, \bar{t}_{j_q} \rangle_R \subseteq R^n$$

of R^n spanned by $\bar{t}_{j_1}, \ldots, \bar{t}_{j_q}$, say. It immediately follows that

$$\langle \bar{s}_{i_1}, \ldots, \bar{s}_{i_p}, \bar{t}_{j_1}, \ldots, \bar{t}_{j_q} \rangle_R = R^n.$$

Write

$$\bar{S}_I = \bar{s}_{i_1} \wedge \cdots \wedge \bar{s}_{i_p} = \sum_{H \in C(n,p)} \det(S_{H,I}) \bar{E}_H;$$

$$\bar{T}_J = \bar{t}_{j_1} \wedge \cdots \wedge \bar{t}_{j_q} = \sum_{K \in C(n,q)} \det(T_{K,J}) \bar{E}_K,$$

where

$$I * J = 1 < 2 < \cdots < n := N.$$

Now denote

$$\det[\bar{S}_I, \bar{T}_J] := \det[\bar{s}_1, \ldots, \bar{s}_p, \bar{t}_1, \ldots, \bar{t}_q]$$

which is an $n \times n$ determinant. Then

$$\bar{S}_I \wedge \bar{T}_J = \mathrm{sign}(\tau_{I,J}) \det[\bar{S}_I, \bar{T}_J] \cdot \bar{E}_{I*J}$$

$$= \sum_{H \in C(n,p), K \in C(n,q)} \mathrm{sign}(\tau_{H,K}) \det(S_{H,I}) \det(T_{K,J}) \bar{E}_{H*K}.$$

For $H \in C(n, p)$, we set $\hat{H} = N \setminus H$, which implies

$$\hat{H} \cap H = \emptyset, \quad \hat{H} \in C(n, q).$$

Thus we get a one to one correspondence between the sets $C(n, p)$ and $C(n, q)$. Note that we also have

$$\dim_R \bigwedge^p M = \#C(n, p) = \binom{n}{p} = \#C(n, q) = \dim_R \bigwedge^q M, \quad M = R^n.$$

It follows from

$$\begin{cases} H * \hat{H} = N = 1 < 2 < \cdots < n, \\ H \cap K \neq \emptyset, \quad \text{if} \quad K \neq \hat{H}, \end{cases}$$

that

$$\overline{S}_I \wedge \overline{T}_J = \sum_{H \in C(n,p), K \in C(n,q)} \text{sign}(\tau_{H,K}) \det(S_{H,I}) \det(T_{K,J}) \overline{E}_{H*K}$$

$$= \left(\sum_{H \in C(n,p)} \text{sign}(\tau_{H,\hat{H}}) \det(S_{H,I}) \det(T_{\hat{H},J}) \right) \overline{E}_{1<2<\cdots<n}.$$

Thus

$$\|\overline{S}_I \wedge \overline{T}_J\|^2 = \left| \sum_{H \in C(n,p)} \text{sign}(\tau_{H,\hat{H}}) \det(S_{H,I}) \det(T_{\hat{H},J}) \right|^2 .$$

We have $T = S^\perp$ which, by Lemma 11.2, implies

$$\|\overline{S}_I \wedge \overline{T}_J\|^2 = \|\overline{S}_I\|^2 \|\overline{T}_J\|^2 = \sum_{H \in C(n,p)} |\det(S_{H,I})|^2 \sum_{H \in C(n,p)} |\det(T_{\hat{H},J})|^2.$$

Hence

$$\left| \sum_{H \in C(n,p)} \text{sign}(\tau_{H,\hat{H}}) \det(S_{H,I}) \det(T_{\hat{H},J}) \right|^2$$

$$= \sum_{H \in C(n,p)} |\det(S_{H,I})|^2 \sum_{H \in C(n,p)} |\det(T_{\hat{H},J})|^2. \quad (11.8)$$

Let us write

$$\begin{cases} \overline{S}_G = \left(\det(S_{H,I}) \right)_{H \in C(n,p)} \in R^{\binom{n}{p}}, \\ \overline{T}_G = \left(\det(T_{\hat{H},J}) \right)_{H \in C(n,p)} \in R^{\binom{n}{p}}, \end{cases}$$

for the Grassmann coordinate vectors of

$$\begin{cases} \overline{s}_{i_1} \wedge \cdots \wedge \overline{s}_{i_p} = \sum_{H \in C(n,p)} \det(S_{H,I}) \overline{E}_H, \\ \overline{t}_{j_1} \wedge \cdots \wedge \overline{t}_{j_q} = \sum_{K \in C(n,q)} \det(T_{K,J}) \overline{E}_K. \end{cases}$$

Then (11.8) may be written as an equality

$$|\overline{S}_G^{\pm} \cdot \overline{T}_G| = \|\overline{S}_G^{\pm}\| \|\overline{T}_G\|, \quad (11.9)$$

where

$$\overline{S_G^{\pm}} := \left(\mathrm{sign}(\tau_{H,\hat{H}}) \overline{\det(S_{H,I})} \right)_{H \in C(n,p)} \in R^{\binom{n}{p}}.$$

12 Rational Subspace

Let K be a field containing the rational number field $\mathbb{Q} \subseteq K$, e.g. $K = \mathbb{R}$ or $K = \mathbb{C}$.

Definition 12.1 Let

$$\overline{a}_1, \ldots, \overline{a}_k \in \mathbb{Q}^n$$

be rational vectors, then the set

$$S = S_k := K\overline{a}_1 + \cdots + K\overline{a}_k,$$

is an k-dimensional rational subspace of K^n, if $\dim_K S = k$.

Lemma 12.2 *Let S_k be a k-dimensional rational subspace of K^n. Then the orthogonal complement*

$$S_k^{\perp} = \{\overline{y} \in K^n | \overline{y} \cdot \overline{x} = 0 \ \forall \overline{x} \in S_k\}$$

is an $(n - k)$-dimensional rational subspace of K^n.

Proof By assumption we have

$$S_k = K\overline{a}_1 + \cdots + K\overline{a}_k, \quad \overline{a}_1, \ldots, \overline{a}_k \in \mathbb{Q}^n, \quad \dim_K S_k = k.$$

Thus

$$V := \mathbb{Q}\overline{a}_1 + \cdots + \mathbb{Q}\overline{a}_k$$

is a k-dimensional subspace of $\mathbb{Q}^n$. The orthogonal complement $V^{\perp}$ of V in $\mathbb{Q}^n$ is a subspace in $\mathbb{Q}^n$ of dimension $\dim_{\mathbb{Q}} V^{\perp} = n - k$. Hence

$$V^{\perp} = \mathbb{Q}\overline{b}_1 + \cdots + \mathbb{Q}\overline{b}_{n-k}, \quad \overline{b}_1, \ldots, \overline{b}_{n-k} \in \mathbb{Q}^n.$$

Further,

$$W := K \otimes_{\mathbb{Q}} V^{\perp} = K\overline{b}_1 + \cdots + K\overline{b}_{n-k}, \quad \overline{b}_1, \ldots, \overline{b}_{n-k} \in \mathbb{Q}^n,$$

is a rational subspace in K^n of dimension $\dim_K W = n - k$ and is the orthogonal complement of S in K^n. $\square$

12.1   Lattices of Subspaces

Lemma 12.3 *Let S_k be a k-dimensional rational subspace of K^n. Then the set*

$$\Lambda(S_k) := S_k \cap \mathbb{Z}^n, \quad rank\ \Lambda(S_k) = k,$$

of integer points in S_k forms a full lattice in the rational subspace S_k.

Let $\bar{l}_1, \ldots, \bar{l}_k \in \mathbb{Z}^n$ be a base of $\Lambda(S_k)$. Then we may write

$$\Lambda(S_k) = \langle \bar{l}_1, \ldots, \bar{l}_k \rangle_\mathbb{Z} = \mathbb{Z}\bar{l}_1 + \cdots + \mathbb{Z}\bar{l}_k \subseteq \mathbb{R}^n.$$

12.2   Height of Rational Subspace

Let S_k be a k-dimensional rational subspace of K^n. Remember that the determinant of $\Lambda(S_k)$ is given by

$$\det(\Lambda(S_k)) := \sqrt{\det[\bar{s}_{i_k} \cdot \bar{s}_{i_l}]_{1 \le k, l \le p}} = \sqrt{\det(L^t L)}, \quad L = [\bar{s}_{i_1}, \ldots, \bar{s}_{i_p}].$$

where the columns of the matrix L are the base vectors $\bar{s}_{i_1}, \ldots, \bar{s}_{i_p}$.

Definition 12.4 Let S_k be a k-dimensional rational subspace of K^n. Then

$$H(S_k) := \det \Lambda(S_k)$$

is the height of S_k. If $k = 0$, then

$$H(S_0) := 1.$$

Vaaler [15] proved the following k-dimensional volume estimate for the intersection of a k-dimensional subspace with the n-dimensional unit cube $\square_n \subseteq \mathbb{R}^n$ of volume $V(\square_n) = 2^n$.

Theorem 12.5 ([15]) *Let $n \in \mathbb{Z}^+$ and let S be a k-dimensional subspace of $\mathbb{R}^n$. Then*

$$\underset{\dim=k}{V}\ (\square_n \cap S_k) \ge 2^k.$$

Lemma 12.6 *Let S_k be a k-dimensional rational subspace of $\mathbb{R}^n$ and let $\lambda_1, \ldots \lambda_k$ be the successive minima of $\square_n$ with respect to $\Lambda = \Lambda(S_k) = S_k \cap \mathbb{Z}^n$, the lattice of integer points in S_k. Then*

$$\lambda_1 \cdots \lambda_k \le \det \Lambda = H(S_k). \tag{12.1}$$

Proof The set $C := \square_n \cap S_k$ is a central symmetric convex body of volume $V(\square_n \cap S_k) \geq 2^k$. Hence, by the Minkowski's second theorem, Theorem 3.6, we get

$$\lambda_1 \cdots \lambda_k 2^k \leq \lambda_1 \cdots \lambda_k V(C) \leq 2^k \det \Lambda,$$

which implies

$$\lambda_1 \cdots \lambda_k \leq \det \Lambda = H(S_k). \quad \square$$

12.3 Primitive Vector

Let $\Lambda(S) = \langle \overline{s}_{i_1}, \ldots, \overline{s}_{i_p} \rangle_{\mathbb{Z}}$ be the integer lattice of a rational subspace $S \subseteq K^n$. Denote by

$$\overline{S}_G = \left(\det(S_{H,I}) \right)_{H \in C(n,p)} \in \mathbb{Z}^{\binom{n}{p}} \tag{12.2}$$

the Grassmann coordinate vector of the exterior product expansion

$$\overline{s}_{i_1} \wedge \cdots \wedge \overline{s}_{i_p} = \sum_{H \in C(n,p)} \det(S_{H,I}) \overline{E}_H,$$

where $I = i_1, \ldots, i_p$ is an increasing $1 \leq p \leq n$-list. We also say that the vector (12.2) is a Grassmann coordinate vector of the integer lattice of a rational subspace S.

Lemma 12.7 (Lemma 5H in [11]) *Let S be a rational subspace in K^n, $\mathbb{Q} \subseteq K$. Then the Grassmann coordinate vector*

$$\overline{S}_G = \left(\det(S_{H,I}) \right)_{H \in C(n,p)} \in \mathbb{Z}^{\binom{n}{p}}$$

is primitive.

Proof Suppose on the contrary that there exists a prime $\ell \in \mathbb{Z}$ such that

$$\overline{S}_G = \left(\det(S_{H,I}) \right)_{H \in C(n,p)} = \ell \cdot \overline{V}, \quad \overline{V} \in \mathbb{Z}^{\binom{n}{p}}.$$

By reduction modulo ℓ we get

$$\overline{S}_G = \overline{0} \in \mathbb{Z}_\ell^{\binom{n}{p}}.$$

Therefore

$$\overline{s}_{i_1} \wedge \cdots \wedge \overline{s}_{i_p} = \overline{0} \in (\mathbb{Z}_\ell)_p^n.$$

By Lemma 10.8 there exists $r_1, \ldots, r_p \in \mathbb{Z}_\ell$, where say $r_1 \neq 0$, such that

$$r_1 \cdot \overline{s}_{i_1} + \cdots + r_p \cdot \overline{s}_{i_p} = \overline{0} \in (\mathbb{Z}_\ell)^n.$$

Hence

$$r_1 \cdot \overline{s}_{i_1} + \cdots + r_p \cdot \overline{s}_{i_p} = \ell \cdot \overline{w}, \quad \overline{w} \in \mathbb{Z}^n,$$

where $r_1, \ldots, r_p \in \mathbb{Z}$ and $\ell \nmid r_1$. Since S is a rational subspace, we find

$$\overline{w} = \frac{r_1}{\ell} \cdot \overline{s}_{i_1} + \cdots + \frac{r_p}{\ell} \cdot \overline{s}_{i_p} \in S,$$

and thus

$$\overline{w} \in S \cap \mathbb{Z}^n = \Lambda(S) = \mathbb{Z}\overline{s}_{i_1} + \cdots + \mathbb{Z}\overline{s}_{i_p}.$$

But $\frac{r_1}{\ell} \notin \mathbb{Z}$, a contradiction. $\square$

By Lemma 12.7 we know that any Grassmann coordinate vector of the integer lattice of a rational subspace is primitive.

13　Height Theorem

Theorem 13.1 (Lemma 4C in [11])

$$H(S^{\perp}) = H(S). \tag{13.1}$$

The result (13.1) is deep.

Proof Let S and T be the subspaces given in (11.6) and (11.7), respectively. Then we define the corresponding lattices

$$\begin{cases} \Lambda(S) := S \cap \mathbb{Z}^n = \langle \overline{ls}_{i_1}, \ldots, \overline{ls}_{i_p} \rangle_{\mathbb{Z}}, \\ \Lambda(T) := T \cap \mathbb{Z}^n = \langle \overline{lt}_{j_1}, \ldots, \overline{lt}_{j_q} \rangle_{\mathbb{Z}}, \end{cases}$$

but for short we write

$$\begin{cases} \Lambda(S) = \langle \overline{s}_{i_1}, \ldots, \overline{s}_{i_p} \rangle_{\mathbb{Z}}, \\ \Lambda(T) = \langle \overline{t}_{j_1}, \ldots, \overline{t}_{j_q} \rangle_{\mathbb{Z}}, \end{cases}$$

where

$$\overline{s}_{i_1}, \ldots, \overline{s}_{i_p}, \overline{t}_{j_1}, \ldots, \overline{t}_{j_q} \in \mathbb{Z}^n.$$

Let $I = i_1, \ldots, i_p$ and $J = j_1, \ldots, j_q$ be increasing lists, where $p, q \in \{1, \ldots, n\}$, then

$$| \det \Lambda(S)|^2 := | \det[\overline{s}_{i_c} \cdot \overline{s}_{i_d}]_{p \times p}|$$

$$\overset{(11.2)}{=} \| \overline{s}_{i_1} \wedge \cdots \wedge \overline{s}_{i_p} \|^2 = \sum_{H \in C(n,p)} | \det(S_{H,I})|^2 \tag{13.2}$$

and, similarly,

$$|\det \Lambda(T)|^2 = \|\bar{t}_{i_1} \wedge \cdots \wedge \bar{t}_{i_q}\|^2 = \sum_{K \in C(n,q)} |\det(T_{K,J})|^2.$$

By (11.9) we have

$$|\overline{S_G^\pm} \cdot \overline{T}_G| = \|\overline{S_G^\pm}\| \|\overline{T}_G\|,$$

Here we use the Cauchy–Schwarz inequality, Lemma 14.6 with an equality, and thus obtain

$$\|\overline{S_G^\pm}\| \overline{T}_G = \pm \|\overline{T}_G\| \overline{S_G^\pm},$$

where $\overline{S_G^\pm}$ and $\overline{T}_G$ are primitive vectors and $\|\overline{S_G^\pm}\|, \|\overline{T}_G\| \in \mathbb{N}$. Hence

$$\|\overline{T}_G\| = \|\overline{S_G^\pm}\|$$

and thus

$$\overline{T}_G = \pm \overline{S_G^\pm}.$$

Further we get

$$\begin{aligned}
|\det \Lambda(S)|^2 &= \|\bar{s}_{i_1} \wedge \cdots \wedge \bar{s}_{i_p}\|^2 \\
&= \sum_{H \in C(n,p)} |\det(S_{H,I})|^2 \\
&= \sum_{H \in C(n,p)} |\text{sign}(\tau_{H,\hat{H}}) \overline{\det(S_{H,I})}|^2 \\
&= \|\overline{S_G^\pm}\|^2 = \|\overline{T}_G\|^2 \\
&= \sum_{K \in C(n,q)} |\det(T_{K,J})|^2 \\
&= \|\bar{t}_{i_1} \wedge \cdots \wedge \bar{t}_{i_q}\|^2 = |\det \Lambda(T)|^2.
\end{aligned}$$

Finally

$$H(S^\perp) := |\det \Lambda(T)| = |\det \Lambda(S)| := H(S). \quad \square$$

In the following chapter we shall give an application of Theorem 13.1. Namely, we shall prove a slightly refined version of Siegel's lemma using Theorem 13.1.

14 Bombieri–Vaaler Version of Siegel's Lemma

In the last part of our lectures we shall present a simplified proof of the Bombieri–Vaaler version of Siegel's lemma, see Bombieri and Vaaler [3]. The proof is based on the second Minkowski's convex body theorem, the height theorem and some careful analysis of primitive Grassmann coordinate vectors.

14.1 Rational Subspaces

So, again, our target is to study integer solutions of the Eq. (6.1). Let us write the Eq. (6.1) as a single matrix equation

$$A\bar{x} = \begin{bmatrix} a_{11} & \cdots & a_{1N} \\ & \cdot & \\ \cdot & \cdot & \cdot \\ & \cdot & \\ a_{M1} & \cdots & a_{MN} \end{bmatrix} \begin{bmatrix} x_1 \\ \cdot \\ \cdot \\ \cdot \\ x_N \end{bmatrix} = 0,$$

where

$$A := [a_{mn}] = \begin{bmatrix} \bar{b}_1^t \\ \cdot \\ \cdot \\ \cdot \\ \bar{b}_M^t \end{bmatrix} \in M_{M \times N}(\mathbb{Z})$$

and $\bar{b}_1^t = (a_{11}, \ldots, a_{1N}), \ldots, \bar{b}_M^t = (a_{M1}, \ldots, a_{MN})$ denote the rows of A.

However, now we assume w.l.o.g. that rank $A = M$. Thus we may consider A as a mapping

$$A : \mathbb{Q}^N \to \mathbb{Q}^M$$

of rank $A = M$. Hence the kernel

$$R := \ker A \subseteq \mathbb{Q}^N, \quad \dim_{\mathbb{Q}} R = N - M,$$

is an $(N - M)$-dimensional rational subspace of $\mathbb{Q}^N$. Further, the orthogonal complement

$$R^\perp \subseteq \mathbb{Q}^N, \quad \dim_{\mathbb{Q}} R^\perp = M$$

is an M-dimensional rational subspace of $\mathbb{Q}^N$.

Next we define the corresponding integer lattices

$$\begin{cases} \Lambda(R) = R \cap \mathbb{Z}^N, & \text{rank } \Lambda(R) = N - M; \\ \Lambda(R^\perp) = R^\perp \cap \mathbb{Z}^N, & \text{rank } \Lambda(R^\perp) = M, \end{cases}$$

where $\Lambda(R^\perp) = \langle \bar{r}_1, \ldots, \bar{r}_M \rangle_{\mathbb{Z}}$ is spanned by integer vectors $\bar{r}_1, \ldots, \bar{r}_M \in \mathbb{Z}^N$. By (13.1) we get

$$H(R) = H(R^\perp) = \det \Lambda(R^\perp) = \sqrt{\det[\bar{r}_m \cdot \bar{r}_n]_{1 \le m,n \le M}}.$$

On the other hand

$$\bar{b}_1, \ldots, \bar{b}_M \in R^\perp \cap \mathbb{Z}^N = \langle \bar{r}_1, \ldots, \bar{r}_M \rangle_{\mathbb{Z}}, \quad \text{rank } \Lambda(R^\perp) = M.$$

Thus we may write $\bar{b}_1, \ldots, \bar{b}_M$ as linear combinations

$$\bar{b}_i = c_{1i}\bar{r}_1 + \cdots + c_{Mi}\bar{r}_M, \quad c_{ji} \in \mathbb{Z}, \quad j, i = 1, \ldots, M.$$

in the base $\bar{r}_1, \ldots, \bar{r}_M$. Also we denote

$$\mathcal{C} := [\bar{c}_1, \ldots, \bar{c}_M] = \begin{bmatrix} c_{1,1} & \cdots & c_{1,M} \\ & \cdot & \\ \cdot & \cdot & \cdot \\ & \cdot & \\ c_{M,1} & \cdots & c_{M,M} \end{bmatrix}.$$

Then

$$\bar{b}_1 \wedge \cdots \wedge \bar{b}_M = \det \mathcal{C} \cdot \bar{r}_1 \wedge \cdots \wedge \bar{r}_M, \quad \det \mathcal{C} \in \mathbb{Z} \setminus \{0\}. \tag{14.1}$$

14.2 Grassmann Coordinates

Next we note that the transpose matrix

$$\mathcal{A}^t = [\bar{b}_1, \ldots, \bar{b}_M] \in M_{N \times M}(\mathbb{Z})$$

of $\mathcal{A}$ is determined by the column vectors $\bar{b}_1, \ldots, \bar{b}_M$. Further, we use the notation

$$\mathcal{R} = [\bar{r}_1, \ldots, \bar{r}_M] \in M_{N \times M}(\mathbb{Z})$$

for the matrix determined by the column vectors $\bar{r}_1, \ldots, \bar{r}_M$. Here the notations $M_1 = M_2 = 1, 2, \ldots, M$ denote increasing lists $1 < 2 < \cdots < M$. By the wedge product expansion rule (10.3) we get

$$\begin{cases} \overline{b}_1 \wedge \cdots \wedge \overline{b}_M = \displaystyle\sum_{H \in C(N,M)} \det(A_{H,M_1})\overline{E}_H, \\ \overline{r}_1 \wedge \cdots \wedge \overline{r}_M = \displaystyle\sum_{K \in C(N,M)} \det(R_{K,M_2})\overline{E}_K, \end{cases}$$

where $\det(A_{H,M_1})$ and $\det(R_{K,M_2})$ are $M \times M$-minors of $\mathcal{A}$ and $\mathcal{R}$, respectively. Also we write

$$\begin{cases} \overline{A}_G := \left(\det(A_{H,M_1})\right)_{H \in C(N,M)} \in \mathbb{Z}^{\binom{N}{M}}, \\ \overline{R}_G := \left(\det(R_{K,M_2})\right)_{K \in C(N,M)} \in \mathbb{Z}^{\binom{N}{M}}, \end{cases}$$

for the Grassmann coordinate vectors of $\overline{b}_1 \wedge \cdots \wedge \overline{b}_M$ and $\overline{r}_1 \wedge \cdots \wedge \overline{r}_M$, respectively.

In the following lemma we give a divisibility relation between the greatest common factor of the minors $\det(A_{H,M_1})$ and $\det \mathcal{C}$.

Lemma 14.1 *Let*

$$D = \gcd_{H \in C(N,M)} \left(\det(A_{H,M_1})\right),$$

then

$$D = |\det \mathcal{C}|. \tag{14.2}$$

Proof By (14.1) we have

$$\overline{A}_G = \det \mathcal{C} \cdot \overline{R}_G, \quad \det \mathcal{C} \in \mathbb{Z} \setminus \{0\},$$

where by Lemma 12.7 the Grassmann coordinate vector

$$\overline{R}_G = \left(\det(R_{K,M_2})\right)_{K \in C(N,M)}$$

of our lattice $\Lambda(R^\perp) = \langle \overline{r}_1, \ldots, \overline{r}_M \rangle_{\mathbb{Z}}$ is primitive. The above immediately imply (14.2). $\square$

Lemma 14.2 *With the above notations we have*

$$\|\overline{r}_1 \wedge \cdots \wedge \overline{r}_M\| \leq D^{-1}\sqrt{\det(\mathcal{A}\mathcal{A}^t)}. \tag{14.3}$$

Proof Here we combine the results (14.1), (14.2) and Laplace identity (11.2). Thus

$$\begin{aligned} D \cdot \|\overline{r}_1 \wedge \cdots \wedge \overline{r}_M\| &\leq |\det \mathcal{C}| \cdot \|\overline{r}_1 \wedge \cdots \wedge \overline{r}_M\| \\ &= \|\overline{b}_1 \wedge \cdots \wedge \overline{b}_M\| \\ &= \sqrt{\det[\overline{b}_m \cdot \overline{b}_n]_{1 \leq m,n \leq M}} = \sqrt{\det(\mathcal{A}\mathcal{A}^t)}. \quad \square \end{aligned}$$

14.3 A Simplified Proof of the Bombieri–Vaaler Version of Siegel's Lemma

Bombieri and Vaaler [3] presented the following improved version of Siegel's lemma.

Theorem 14.3 *Let*
$$\mathcal{A} \in M_{M \times N}(\mathbb{Z}), \quad rank\, \mathcal{A} = M$$

Then the equation
$$\mathcal{A}\bar{x} = \bar{0}$$

has $N - M$ linearly independent integer solutions $\bar{x}_1, \ldots, \bar{x}_{N-M} \in \mathbb{Z}^N \setminus \{\bar{0}\}$ such that
$$\|\bar{x}_1\|_\infty \cdots \|\bar{x}_{N-M}\|_\infty \leq D^{-1}\sqrt{\det(\mathcal{A}\mathcal{A}^t)}, \tag{14.4}$$

where D is the greatest common divisor of all $M \times M$ minors of $\mathcal{A}$.

Proof Recall that R is an $(N - M)$-dimensional rational subspace in $\mathbb{Q}^N$. Let $\lambda_1, \ldots, \lambda_{N-M}$ be the successive minima of the unit cube $\square_N$ with respect to $\Lambda = \Lambda(R) = R \cap \mathbb{Z}^N$, the lattice of integer points in R. Then Lemma 12.6 applied to the rational subspace R with the height Theorem 13.1 gives
$$\lambda_1 \cdots \lambda_{N-M} \leq H(R) = H(R^\perp) = \det \Lambda(R^\perp). \tag{14.5}$$

By Example 3.5 there exist $N - M$ linearly independent integer vectors $\bar{x}_1, \ldots \bar{x}_{N-M} \in \mathbb{Z}^N \setminus \{\bar{0}\}$ from R such that
$$\|\bar{x}_1\|_\infty \leq \lambda_1, \ldots, \|\bar{x}_{N-M}\|_\infty \leq \lambda_{N-M}. \tag{14.6}$$

Hence, by (14.5), (14.6) and (14.3), it follows that
$$\begin{aligned}
\|\bar{x}_1\|_\infty \cdots \|\bar{x}_{N-M}\|_\infty &\leq \lambda_1 \cdots \lambda_{N-M} \\
&\leq H(R) = H(R^\perp) \\
&= \det \Lambda(R^\perp) \\
&= \sqrt{\det[\bar{r}_m \cdot \bar{r}_n]_{1 \leq m,n \leq M}} \\
&= \|\bar{r}_1 \wedge \cdots \wedge \bar{r}_M\| \leq D^{-1}\sqrt{\det(\mathcal{A}\mathcal{A}^t)}. \quad \square
\end{aligned}$$

Corollary 14.4 *In Theorem 14.3 the estimate* (14.4) *can be replaced by (possibly a weaker estimate)*
$$\prod_{k=1}^{N-M} \|\bar{x}_k\|_\infty \leq \prod_{m=1}^{M} \|\bar{b}_m\|_2. \tag{14.7}$$

There are several variants of Siegel's lemma over $\mathbb{Q}$ which follow from Corollary 14.4. In the following corollary we present two of them.

Corollary 14.5 *In Theorem 14.3 the estimate* (14.4) *can be replaced by (possibly weaker estimates)*

$$\|\overline{x}_1\|_\infty \leq \left(\prod_{m=1}^{M} \|\overline{b}_m\|_1 \right)^{\frac{1}{N-M}} \leq \left(N \max_{1 \leq m,n \leq N} |a_{m,n}| \right)^{\frac{M}{N-M}}. \tag{14.8}$$

In Corollary 14.5 the first upper bound is exactly the upper bound in the estimate (6.4).

The above corollaries follow from estimate (14.12) for the well-known Gram determinant $\det[\overline{b}_m \cdot \overline{b}_n]_{1 \leq m,n \leq M}$. Notice that our Gram determinant is positive.

In fact, Bombieri and Vaaler [3] proved more general results over algebraic numbers by using geometry of numbers over the adéles.

Appendix

Cauchy–Schwarz Inequality

Now we suppose that R is an integral domain with absolute value and the norm of $\overline{a} \in R^N$ is defined by

$$\|\overline{a}\| = \sqrt{\overline{a} \cdot \overline{a}}$$

via the inner product $\cdot$ in R^N.

Lemma 14.6 (Cauchy–Schwarz inequality)

$$|\overline{a} \cdot \overline{b}| \leq \|\overline{a}\| \|\overline{b}\| \quad \text{for all} \quad \overline{a}, \overline{b} \in R^N. \tag{14.9}$$

The equality

$$|\overline{a} \cdot \overline{b}| = \|\overline{a}\| \|\overline{b}\|$$

implies

$$\|\overline{b}\| \overline{a} = \omega \|\overline{a}\| \overline{b}, \quad |\omega| = 1, \ \omega \in R. \tag{14.10}$$

In particular, when $R = \mathbb{Z}$, then $\omega = \pm 1$.

Proof Write

$$\overline{w} := \|\overline{b}\|^2 \overline{a} - (\overline{a} \cdot \overline{b}) \overline{b}.$$

Then

$$\overline{w} \cdot \overline{b} = 0$$

and consequently

$$\|\bar{b}\|^4 \|\bar{a}\|^2 = \|\bar{w}\|^2 + |\bar{a} \cdot \bar{b}|^2 \|\bar{b}\|^2 \geq |\bar{a} \cdot \bar{b}|^2 \|\bar{b}\|^2 \qquad (14.11)$$

which implies (14.9).

Suppose now

$$|\bar{a} \cdot \bar{b}| = \|\bar{a}\| \|\bar{b}\| \quad \Leftrightarrow \quad \bar{a} \cdot \bar{b} = \omega \|\bar{a}\| \|\bar{b}\|, \quad |\omega| = 1, \ \omega \in R.$$

Then (14.11) reads

$$\|\bar{b}\|^4 \|\bar{a}\|^2 = \|\bar{w}\|^2 + \|\bar{a}\|^2 \|\bar{b}\|^4.$$

Hence,

$$\|\bar{w}\|^2 = 0 \quad \Rightarrow \quad \bar{w} = \|\bar{b}\|^2 \bar{a} - (\bar{a} \cdot \bar{b})\bar{b} = \bar{0}$$

$$\Rightarrow \quad \|\bar{b}\|^2 \bar{a} = (\bar{a} \cdot \bar{b})\bar{b} = \omega \|\bar{a}\| \|\bar{b}\| \bar{b},$$

which proves (14.10). $\square$

14.4   Gram Determinant

The Gram determinant $\det[\bar{b}_m \cdot \bar{b}_n]_{1 \leq m, n \leq M}$ satisfies the following estimates

$$\sqrt{\det(\mathcal{A}\mathcal{A}^t)} = \sqrt{\det[\bar{b}_m \cdot \bar{b}_n]_{1 \leq m, n \leq M}}$$

$$\leq \sqrt{\prod_{m=1}^{M} \bar{b}_m \cdot \bar{b}_m} = \prod_{m=1}^{M} \|\bar{b}_m\|_2$$

$$\leq \prod_{m=1}^{M} \|\bar{b}_m\|_1 \leq \prod_{m=1}^{M} N \max_{1 \leq n \leq N} |a_{m,n}|$$

$$\leq \left(N \max_{1 \leq m, n \leq N} |a_{m,n}| \right)^M. \qquad (14.12)$$

References

1. A. Baker, On some Diophantine inequalities involving the exponential function. Canad. J. Math. **17**, 616–626 (1965)
2. A. Baker, *Transcendental Number Theory* (Cambridge University Press, Cambridge, 1975)
3. E. Bombieri, J.D. Vaaler, On Siegel's lemma. Invent. Math. **73**, 11–32 (1983)
4. E. Bombieri et al., *Recent Progress in Analytic Number Theory*, vol. 2, On *G*-functions (Academic Press, London-New York, 1981), pp. 1–67. (Durham, 1979)

5. J.W.S. Cassels, *An introduction to the geometry of numbers*. Corrected reprint of the Classics in Mathematics, 1971 edition (Springer, Berlin, 1997)
6. A.-M. Ernvall-Hytönen, K. Leppälä, T. Matala-aho, An explicit Baker type lower bound of exponential values. Proc. Roy. Soc. Edinburgh Sect. A **145**, 1153–1182 (2015)
7. K. Mahler, On a paper by A. Baker on the approximation of rational powers of *e*. Acta Arith. **27**, 61–87 (1975)
8. T. Matala-aho, On Baker type lower bounds for linear forms. Acta Arith. **172**, 305–323 (2016)
9. J.J. Rotman, *Advanced Modern Algebra* (Pearson, New York, 2002)
10. W.M. Schmidt, *Diophantine Approximation*, vol. 785, Lecture Notes in Mathematics (Springer, Berlin, 1980)
11. W.M. Schmidt, *Diophantine approximations and Diophantine equations*, vol. 1467, Lecture Notes in Mathematics (Springer, Berlin, 1991)
12. A.B. Shidlovskii, *Transcendental numbers, de Gruyter Studies in Mathematics 12* (Walter de Gruyter and Co., Berlin, 1989)
13. C.L. Siegel, *Transcendental Numbers*, vol. 16, Annals of Mathematics Studies (Princeton university press, Princeton, 1949)
14. J. Steuding, *Diophantine analysis* (Chapman & Hall/CRC, Boca Baton, 2005)
15. J.D. Vaaler, A geometric inequality with applications to linear forms. Pacific J. Math. **83**, 543–553 (1979)

Historical Face of Number Theory(ists) at the Turn of the 19th Century

Nicola Oswald

2000 Mathematics Subject Classification 01A60 · 01A70 · 11J70 · 37A05

Some Introductory Words

Of course, it is impossible to give here a complete overview of the many aspects of number theory and its protagonists. Consequently, we focus mainly on one face of its multifaceted history. The concrete program consists of the following four topics:

- **Meeting the Hurwitz brothers.**
  The brothers Julius (1857–1919) and Adolf Hurwitz (1859–1919) were both gifted with a smart intellect and great curiosity in science. Already during their school-days the two of them became acquainted with mathematical problems and both started studies in mathematics. So far nothing extraordinary. While the younger brother turned out to be extremely successful in his research, the elder brother and his work, however, seem to be almost forgotten.
- **Adolf Hurwitz's mathematical diaries: an example of understanding mathematics on behalf of historical documents.**
  The mathematical notebooks of Adolf Hurwitz are stored in the library of the ETH Zurich [21]. Alongside a variety of publications, the zealous Hurwitz wrote in a meticulous manner mathematical diaries from March 1882 to September 1919. Mostly with an accurate writing and an impressive precision Adolf Hurwitz worked on proofs of colleagues, made notes for future dissertation topics and developed his own approaches to various mathematical problems. We will focus on a certain entry in which Adolf Hurwitz refers to his elder brother's dissertation. How do his notes resemble Julius Hurwitz's mathematical results?

N. Oswald (✉)
Faculty of Mathematics and Natural Sciences, University of Wuppertal,
GaußStr. 20, 42119 Wuppertal, Germany
e-mail: oswald@uni-wuppertal.de

© Springer International Publishing AG 2016
J. Steuding (ed.), *Diophantine Analysis*, Trends in Mathematics,
DOI 10.1007/978-3-319-48817-2_4

- **A fruitful friendship: Adolf Hurwitz and David Hilbert.**
  We give a sketch of an approach to the relationship of two great mathematicians on behalf of methods from history. For this purpose we define the mathematical diaries as main corpus of the examination. Those highlight that the lifelong friendship of Adolf Hurwitz with his former student and later colleague David Hilbert was not only exceptionally fruitful, but also rather interesting. We shall consider the question whether their relation had undergone a certain change between 1884 and 1919. It appeared that within this period the famous Hilbert completely emancipated from his teacher Hurwitz.
- **Julius Hurwitz and an ergodic theoretical view on his complex continued fraction.**
  In his dissertation [30] Julius Hurwitz defined a certain continued fraction for complex numbers. We consider his approach and derive an ergodic theoretical result. More precisely, on behalf of characteristics which were worked out by Shigeru Tanaka in [46] we shall sketch a proof of the analogue of the so called Doeblin-Lenstra Conjecture for Julius Hurwitz's continued fractions. Interestingly, the Japanese mathematician published his results nearly one hundred years after Julius, probably without knowing the old dissertation.

These notes deal with a combination of Diophantine number theory and history of mathematics. On the first glimpse this might be unexpected, however, at a second glance this can allow rather fruitful points of view. Doing research in mathematics, it turns out to be very helpful to be aware of the early considerations and the development of the corresponding subject. We want to work out how such an approach could look like.

1 Meeting the Hurwitz Brothers

Reading the title of this section, the reader interested in Diophantine analysis might ask the following question:

> Why is it interesting to know more about the Hurwitz brothers at all?

Reading the name Hurwitz, probably most mathematicians think of Adolf Hurwitz who was a great scientist in a broad range of mathematical disciplines. Here we are mostly concerned about his work on complex continued fractions. Investigating the scientific contributions of "Hurwitz", one probably notices an article entitled "*On a certain kind of the continued fraction expansion of complex values*"[1] [30] written by J. Hurwitz. This can be a starting point of becoming curious: is there a typing error? If Adolf Hurwitz is not the author, who else? In the following we give an insight into the work with archive sources by reconstructing the biography of this other Hurwitz (Fig. 1).

---

[1]"*Ueber eine besondere Art der Kettenbruch-Entwickelung complexer Grössen*".

Fig. 1 Portraits of Adolf Hurwitz (1859–1919) (*on the left*) and Julius Hurwitz (1857–1919), taken from Riesz's register in *Acta Mathematica* from 1913 [42]

The fact that the two mathematicians were relatives becomes clear rather quickly on behalf of documents stored in the municipal archive of Adolf Hurwitz's birthplace. According to them, Julius (born 14 July 1857) and Adolf Hurwitz (born 26 March 1859) were brothers and grew up in Hildesheim, a tranquil town near Hanover. They were born into a Jewish merchant family and after the early death of their mother Elise Wertheim-Hurwitz in 1862, their father Salomon Hurwitz (1813–1885) raised them and their elder brother Max (1855–1910) with the help of his sister Rosette Hurwitz. The latter information comes from a biographical dossier of Adolf's wife Ida Samuel-Hurwitz which can be found in the university archive of the ETH in Zurich [45]. Following her description, Salomon was a hard working man, being the manager of his own local textile manufactory, and even though the family lived in modest surroundings, he was willing to provide his sons with a good education. In their youth nothing indicated how different their lives would evolve. We can read that both, Adolf as well as Julius, were sent to the Realgymnasium "Adreanum",[2] both sons "showed a particular talent for mathematics, which was documented at a very early stage [...]" [45, p. 4] and both received extra lessons by their teacher Hermann Caesar Hannibal Schubert (1848–1911) on Sunday afternoons. Since Schubert spent only a few years in Hildesheim before he moved further to Hamburg [5, 6] where he became a famous and influential geometer, the brothers were very lucky that their times overlapped. In one of Julius's school certificates, also stored in the municipal archive of Hildesheim, the note "Julius visited a tavern without permission" indicates that the elder brother seemed to have been a bit more open for distraction. Nevertheless, "Schubert took the trouble to visit the father to convince him to let both sons take up the study of mathematics [...]. [Salomon] talked to a wealthy and childless

---

[2]At that time this was a new kind of highschool with a focus on mathematics and natural sciences.

friend, E. Edwards, who offered to bear the costs of the studies for one of the sons. Dr. Schubert selected Adolf." [45, p. 4] Consequently, "Max and Julius had to become business men after the graduation of the upper secondary. The pecuniary situation as well as the future plans of the father for his sons did not allow to even consider an academic profession."[3] [45, p. 4] This decision separated at least the professional lives of the brothers fundamentally. The younger enrolled at the University of Munich and obtained a doctorate supervised by Felix Klein (1849–1925) in Leipzig. This was the beginning of a meteoric career: Adolf Hurwitz got to know the academic society in Berlin and, with the support of Karl Weierstrass (1815–1897) and Hermann Amandus Schwarz (1843–1921), he did his habilitation in Göttingen. This enabled him to receive a professorship at the University of Königsberg in 1884 on promotion of Ferdinand von Lindemann (1852–1939), another pupil of Felix Klein. What is particularly noteworthy here is that Adolf Hurwitz got to know different schools of mathematics, was supported by some of the most influencial mathematcians of his time, and was on his way to become an integral part of the mathematical community himself.[4]

This contrasts with a completely different career for Julius. The elder brother had to postpone his academic ambitions, left school without a general higher education entrance qualification and started a banking apprenticeship in Nordhausen, a small village in what is nowadays Thuringia in Germany.[5] Later Julius took over the banking business from his uncle Adolph M. Wertheimer in Hanover. However, there are some hints that Julius never broke with his mathematical interests. An evidence therefore can be found in a letter from the collected letter exchange of Adolf Hurwitz at the university archive in Göttingen. Here the Hanoveranian mathematician Hans von Mangoldt (1854–1925) [48] mentioned the elder brother, which shows that Julius Hurwitz maintained active contact with mathematicians from his place of residence. Correspondingly, his sister-in-law remembered that he never got accustomed with his profession as a banker. She wrote that he "[...] felt uncomfortable in this business. Hence, [...] Julius left in order to return to school at the age of 33 and passed the school examination [...]" [45, p. 8]. More precisely, Julius Hurwitz attended a Realgymnasium in Quakenbrück, a small town in northern Germany. And again, we can still find his school leaving certificate from September 9, 1890 in the archive of the University of Halle.[6] It is rather impressive to read: he not only had to complete an exam in mathematics but also in various subjects like geography, French and even gymnastics and drawing. Furthermore, he was assessed in the category of "moral behavior", which was considered to be excellent in his case. A glance at the list of graduates [1, p. 146] of the school illustrates how unusual and demanding Julius

---

[3]"Max und Julius mussten sich, nach Absolvierung der Ober-Secunda, dem Kaufmannsstand widmen. Sowohl die pekunäre Lage, wie auch die Zukunftspläne des Vaters für seine Söhne liessen einen gelehrten Beruf gar nicht in Frage kommen."

[4]For a more detailed biography see [38].

[5]Side remark: Nordhausen was also the place where the first congress of vegetarians had taken place in 1869.

[6]Archive of Halle University, Rep. 21 Nr. 162.

Hurwitz's late school days must have been. Between 1885 and 1895 the average age of graduates was twenty, around thirteen years younger than him and, furthermore, less than ten students finished school per year. We may conclude that Julius must have had a big motivation to take the risk to change his life so drastically. Probably a certain financial independence arising from the banking business of his uncle helped him with his decision. Another reason could have been the good integration of his brother in Königsberg. Adolf Hurwitz had the great luck to have Hermann Minkowski (1864–1909) and David Hilbert (1862–1943) as students, two extremely talented mathematicians who later became great personalities in mathematics of their generation.

Having graduated from school, Julius Hurwitz moved to his brother and attended lectures in Königsberg. It is known that he edited a variety of lecture notes not only from his brother, but also, among others, from Viktor Eberhard in 1890/91 [28] and David Hilbert in 1892 [29]. On the first page of the second one can find Julius' handwriting: "reviewed by Dr. Hilbert a. equipped with his personal sidenotes".[7] In general, we may assume that Julius Hurwitz was nicely integrated into the academic circle of mathematicians in Königsberg, even though there had always been a strong dependence of his successful brother. In consequence it is not surprising that when Adolf Hurwitz was called for a full professorship to the ETH Zurich in 1892, Ida noted "[Also] his brother Julius soon followed him to Zurich [...]", and she continued: "[to Zurich,] where he wrote his doctoral thesis, for which he had received the subject from his brother." [45, p. 9] In fact, it was not least a consequence of anti-Semitic university-political decisions to Adolf Hurwitz's disadvantage why he moved to Switzerland. There are documents which show that he was considered for positions in Rostock and Göttingen, which he could not receive being a Jew.[8] At that time, professors of the ETH did not have the right to award doctorates and there was consequently not an official option for Julius Hurwitz to proceed at the polytechnic. However, it seems that once more he took a benefit from the good integration of his brother in the mathematical community. In 1893 Julius officially enrolled at a German university, namely the University of Halle-Wittenberg. There he wrote his dissertation under the supervision of Albert Wangerin (1844–1933), who seems to have been a very open-minded teacher. At the end of his career he had supervised the impressive number of 53 doctoral candidates on various topics.[9] In his report on Julius' dissertation Wangerin declared, "J. Hurwitz examines a certain kind of continued fraction expansion of complex numbers following work by his brother, Prof. A. Hurwitz in Zurich, published in Acta Mathematica, volume XI." [49]

After his doctorate, Julius Hurwitz worked for a few years at the University of Basel, however, he only published one more scientific work [31], which included and, in a certain way, extended his dissertation. His brother Adolf Hurwitz stayed in Zurich for the rest of his life.

---

[7]"von Dr. Hilbert durchgesehen u. mit eigenhändigen Randbemerkungen versehen.", archive ETH, HS 582: 154.

[8]For more information see [43, 44].

[9]A list can be found on www.mathematikuni-halle.de/history.

2   Adolf Hurwitz's Mathematical Diaries: An Example of Understanding Mathematics Development on the Basis of Historical Documents

> Since his habilitation in 1882, Hurwitz took notes of everything he spent time on with uninterrupted regularity and in this way he left a series of 31 diaries, which provide a true view of his constantly progressive development and at the same time they are a rich treasure trove for interesting and further examination appropriate thoughts and problems."[10] [17, p. 166]

With David Hilbert's description, taken from his commemorative speech, he provided a good picture of Adolf Hurwitz's mathematical diaries. Those notebooks are filled with interesting problems and inspiring mathematical ideas. After Adolf Hurwitz's death they were reviewed and registered in an additional notebook [21, No. 32] by his confidant and colleague George Pólya, who considered Hurwitz as "colleague who he felt influenced the most." [39, p. 25]

In his ninth diary we can find an entry entitled "Concerning Julius' work." [21, No. 9, pp. 94] which refers to his brother's dissertation on a certain complex continued fraction. In fact, on the very first page of his work Julius Hurwitz wrote that "the thesis follows in aim and method two publications due to Mr. A. Hurwitz to whom I owe the encouragement for this investigation.",[11] which emphasizes that Adolf supported him at certain points. In the two mentioned publications [22, 23] Adolf Hurwitz presented a complex continued fraction allowing all Gaussian integers as possible partial quotients. Interestingly, it is also Adolf Hurwitz who wrote a review in Zentralblatt [24] about Julius' doctorate; it starts as follows:

> The complex plane may be tiled by straight lines $x + y = v$, $x - y = v$ into infinitely many squares, where v ranges through all positive and negative odd integers. The centers of the squares are complex integers divisible by $1 + i$. For an arbitrary complex number x, one may develop the sequence of equations
>
> $$(1) \qquad x = a - \frac{1}{x_1}, \qquad x_1 = a_1 - \frac{1}{x_2}, \qquad \ldots, \qquad x_n = a_n - \frac{1}{x_{n+1}}, \qquad \ldots$$
>
> following the rule that in general a_i is the center of the square which contains x_i. In the case when x_i is lying on the boundary of a square, some further rule has to be applied which we ignore here for the sake of brevity. For x the sequence of equations (1) leads to a continued fraction expansion $x = (a, a_1, \ldots, a_n, x_{n+1})$ which further investigation is the topic of this work.[12]

---

[10]"Hurwitz hat seit seiner Habilitation 1882 in ununterbrochenender Regelmäßigkeit von allem, was ihn wissenschaftlich beschäftigte, Aufzeichnungen gemacht und auf diese Weise eine Serie von 31 Tagebüchern hinterlassen, die ein getreues Bild seiner beständig fortschreitenden Entwicklung geben und zugleich eine reiche Fundgrube für interessante und zur weiteren Bearbeitung geeignete Gedanken und Probleme sind."

[11]"Die Arbeit schliesst sich, nach Ziel und Methode, eng an die nachstehend genannten zwei Abhandlungen des Herrn A. Hurwitz an, dem ich auch die Anregung zu dieser Untersuchung verdanke."

[12]"Die complexe Zahlenebene werde durch die Geraden $x + y = v, x - y = v$, wo ? alle positiven und negativen ungeraden ganzen Zahlen durchläuft, in unendlich viele Quadrate eingeteilt. Die

On the first view Julius' expansion could be mistaken as a specification of the general complex continued fraction investigated by his younger brother. The key difference is a modification of the set of possible partial quotients. Julius restricted this set to the ideal generated by $\alpha := 1 + i$. It shall be noticed that 1 is not an element of this ideal, hence condition iii) of Adolf Hurwitz's setting for his "system" S of partial quotients is not fulfilled (see [22]). Julius' approach results in a thinner lattice in $\mathbb{C}$. This leads to another tiling of the complex plane and enables consequently the definition of a "nearest" partial quotient $a_n \in (\alpha) = (1 + i)\mathbb{Z}[i]$ to each complex number $z \in \mathbb{C}$. What turns up is an expansion of a complex continued fraction

$$z = a_0 - \cfrac{1}{a_1 - \cfrac{1}{a_2 - \cfrac{1}{a_3 + T^3 z}}},$$

where each iteration is determined by a mapping T, which will be defined in Sect. 4. A specific characteristic of this continued fraction is the admissibility of sequences of partial quotients:

What kind of partial quotients can occur?

Adolf Hurwitz's diary entry from November 04, 1894, deals with exactly this question. We find several pages filled with calculations on partial quotients from complex continued fractions and a nice result (Fig. 2).

When compared with a certain paragraph of Julius Hurwitz's doctoral thesis on page 12, we recognize a definite similarity:

We translate Hurwitz's consequences concerning so called not admissible sequences of partial quotients:

Lemma 2.1 ([30])
The following rules for consecutive partial quotients hold:
If a_r is of the type $1 + i$, then a_{r+1} is not of type $2, 1 - i$ or $-2i$.
If a_r is of the type $-1 + i$, then a_{r+1} is not of type $-2i, -1 - i$ or -2.
If a_r is of the type $-1 - i$, then a_{r+1} is not of type $-2, -1 + i$ or $2i$.
If a_r is of the type $1 - i$, then a_{r+1} is not of type $2i, 1 + i$ or 2.

(Footnote 12 continued)
Mittelpunkte dieser Quadrate werden durch die durch 1+i teilbaren ganzen complexen Zahlen besetzt. Wenn nun x eine beliebige complexe Zahl ist, so bilde man die Gleichungskette:

$$(1) \qquad x = a - \frac{1}{x_1}, \qquad x_1 = a_1 - \frac{1}{x_2}, \qquad \ldots, \qquad x_n = a_n - \frac{1}{x_{n+1}}, \qquad \ldots$$

nach der Massgabe, dass allgemein a_i den Mittelpunkt desjenigen Quadrates bezeichnet, in welches der Punkt x_i hineinfällt. Dabei sind noch bezüglich des Falles, wo x_i auf den Rand eines Quadrates fällt, besondere Festsetzungen getroffen, die wir der Kürze halber übergehen. Durch die Gleichungskette (1) wird nun für x eine bestimmte Kettenbruchentwickelung $x = (a, a_1, \ldots, a_n, x_{n+1})$ gegeben, deren nähere Untersuchung der Gegenstand der Arbeit ist."

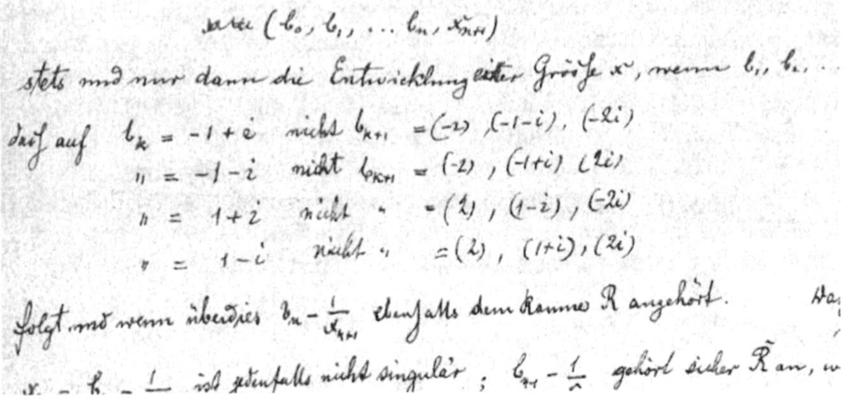

... $(b_0, b_1, \ldots b_n, x_{n+1})$

stets und nur dann die Entwicklung erster Größe x, wenn $b_1, b_2 \ldots$

daß auf $b_k = -1+i$ nicht $b_{k+1} = (-2), (-1-i), (-2i)$

" $= -1-i$ nicht $b_{k+1} = (-2), (-1+i), (2i)$

" $= 1+i$ nicht " $= (2), (1-i), (-2i)$

" $= 1-i$ nicht " $= (2), (1+i), (2i)$

folgt und wenn überdies $x_n - \frac{1}{x_{n+1}}$ ebenfalls dem Raume R angehört. Da

... ist jedenfalls nicht singulär; $b_{n+1} - \frac{1}{x}$ gehört sicher $\bar{R}$ an, ...

Fig. 2 Excerpt from Adolf Hurwitz's mathematical diary entry concerning Julius' work. [21, No. 9, pp. 94]

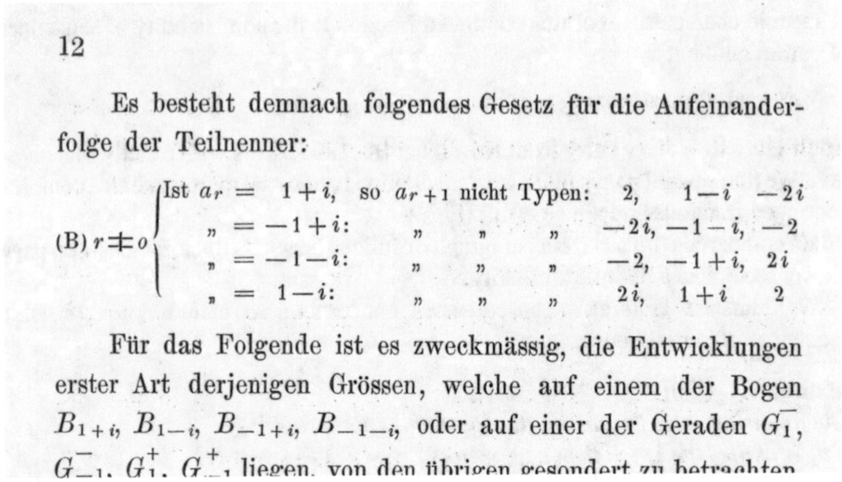

12

Es besteht demnach folgendes Gesetz für die Aufeinanderfolge der Teilnenner:

(B) $r \neq 0$
Ist $a_r = 1+i$, so a_{r+1} nicht Typen: $2,\ 1-i\ -2i$
" $= -1+i$: " " " $-2i,\ -1-i,\ -2$
" $= -1-i$: " " " $-2,\ -1+i,\ 2i$
" $= 1-i$: " " " $2i,\ 1+i\ \ 2$

Für das Folgende ist es zweckmässig, die Entwicklungen erster Art derjenigen Grössen, welche auf einem der Bogen B_{1+i}, B_{1-i}, B_{-1+i}, B_{-1-i}, oder auf einer der Geraden G_1^-, G_{-1}^-, G_1^+, G_{-1}^+ liegen, von den übrigen gesondert zu betrachten

Fig. 3 Excerpt from Julius Hurwitz's doctoral thesis [30, p. 12]

In the following we use Julius Hurwitz's notation (see Fig. 3): firstly, he denotes the half circle in the complex plane through the points i, $1+i$ and 1 by the symbol B_{1+i}. Accordingly, he defines B_{1-i}, B_{-1+i}, and B_{-1-i} in a similar way as half circles with the end points 1 and $-i$, -1 and i, and -1 and $-i$, respectively. Furthermore, he designates the four boundary lines of the fundamental domain. The symbol G_1^- represents the line in the third quadrant, G_{-1}^- the line in the second quadrant, G_1^+ the line in the first and G_1^+ the line in the fourth quadrant.

Certainly, Adolf Hurwitz's notes indicate his direct help and influence on his brother's work. However, here we want to focus on their mathematical result and illustrate their approach to define which successor a certain partial quotients can not

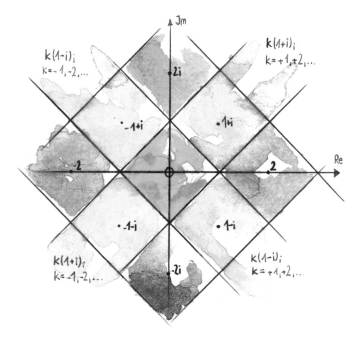

Fig. 4 Illustration, taken from [36], of the types of partial quotients and the associated complete tiling of the complex plane

have. In his table, Julius separated partial quotients into those of type $\pm 1 \pm i$ and those of type ± 2, respectively $\pm i2$. First we need to comprehend this notation. Partial quotients are said to be of *type* $1 + i$ if they are of the form $k(1 + i)$; $k = +1, +2, \ldots$, they are of *type* $1 - i$ if they are of the form $k(1 - i)$; $k = +1, +2, \ldots$, they are of *type* $-1 + i$ if they are of the form $k(1 - i)$; $k = -1, -2, \ldots$, and they are of *type* $-1 - i$ if they are of the form $k(1 + i)$; $k = -1, -2, \ldots$. Furthermore, partial quotients are of *type* 2 when they are located in between the angle bisector of the first and the fourth quadrant, -2 when they are located between the second and third quadrant, and so on.

Notice that this typification provides a complete tiling of the complex plane (see Fig. 4). Now we want to comprehend Hurwitz's line of reasoning. Therefore we pick out the first case of Julius' Lemma 2.1: *If a partial quotient equals $a_n = 1 + i$, then the next partial quotient a_{n+1} is different from* 2, $1 - i$, $-2i$.

The geometrical approach of the Hurwitz brothers to prove this statement can be illustrated in three steps, based on the equation

$$z_n = a_n - \frac{1}{z_{n+1}}.$$

Fig. 5 Excerpt of the
workshop notes illustrating
the inversion

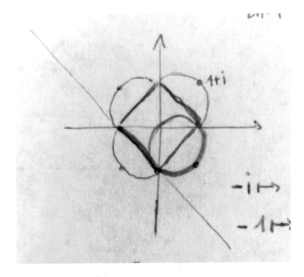

First we take a closer look at the inversion

$$I : X \to X^{-1}, \quad z_{n-1} \mapsto -\frac{1}{z_{n-1}} = z_n,$$

taking place in each iteration of the continued fraction algorithm. Here an element $z_{n-1} \in X$ of the fundamental domain is mapped by a Möbius transformation to the negative of its reciprocal (Fig. 5).

The boundaries of X are represented by the line segments G_1^-, G_{-1}^-, G_1^+ and G_{-1}^+ in the four quadrants. Those are mapped to arcs under the transformation I, which form an inner boundary of the set of reciprocals X^{-1}. We illustrate this by the example of the line G_1^- in the third quadrant having end points -1 and $-i$. Those are transformed as follows:

$$-1 \mapsto -\frac{1}{-1} = 1 \text{ and } -i \mapsto -\frac{1}{-i} = -i.$$

The arc arising from G_1^- is the half circle between the points $-i$ and 1, which does not pass through the origin. Merging all four half circles $B_{\pm 1 \pm i}$ through the points $\pm 1 \pm i$, arising from the inversion $I(G_{\pm 1}^{\pm})$, one can see that the set of reciprocals is given by all complex numbers excluding a flower shaped area F located around the origin: $X^{-1} = \mathbb{C} \setminus F$.

Now we focus on the above stated case $a_n = 1 + i$, which implies that z_n is element of the fundamental domain shifted around $1 + i$. We furthermore know that z_n is not an element of F. The elementary translation

Fig. 6 Illustration of the
translation

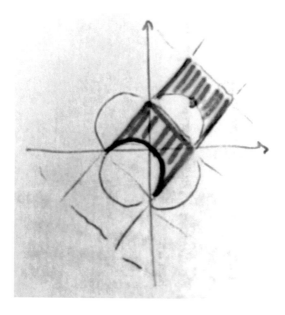

$$z_n \mapsto z_n - a_n = \frac{1}{z_{n+1}}$$

performs a shift back to the fundamental domain X (see Fig. 6).

Notice that the domain of possible reciprocals $\frac{1}{z_{n+1}}$ in X is bounded by G_{-1}^-, G_1^+, G_{-1}^+ and the half circle through the points -1, $-i$ and zero. Since this arc is the continuation of B_{-1-i}, we denote it as B'_{-1-i} in the sequel.

Now we complete our considerations by a second inversion performing

$$\frac{1}{z_{n+1}} \mapsto -z_{n+1}.$$

To figure out the actually possible image domain, we need to focus on the transformation of B'_{-1-i}. We know

$$-1 \mapsto 1 \text{ and } -i \mapsto -i.$$

Furthermore, the origin is mapped to infinity. This leads to a more limited set of possible z_{n+1}, excluding not only F but also that part of the Gaussian complex plane which is bounded by the straight line through $-i$ and 1 (see Fig. 7).

Taking into account Julius Hurwitz's typification of partial quotients, we come to the same conclusion as he and his brother did: a_{n+1} can not be of the type 2, $-2i$ or $1 - i$.

The diary entry of Adolf Hurwitz provides a nice example of how a mathematical result and its reception can evolve. Based on the model of his younger brother's

Fig. 7 Illustration of the
second inversion

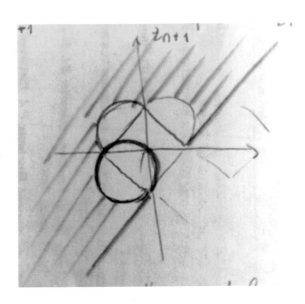

approach, Julius Hurwitz's line of argumentation is based on examining arcs occuring by Möbius transformations. Therefore, he investigated a number of cases for all types of partial quotients in a similar manner as the analyzed example above. Thanks to Adolf Hurwitz's notes we can comprehend the origins of Julius' ideas.

3   A Fruitful Friendship: Adolf Hurwitz and David Hilbert

Examination of Adolf Hurwitz's estate furthermore highlights that his lifelong friendship to his former student and later colleague David Hilbert was not only exceptionally fruitful, but also rather interesting. It seems that their relation had undergone a certain change between 1884 and 1919. In this section we try to approach to a research question from history of mathematics:

> Can we identify indicators marking when the student–teacher relation transformed to a collegial relation?

Hereby, it is important to make clear what "corpus" we use, which means, what documents we take into account. The following analysis is mainly based on the mentioned Hurwitz estate and its comparison to additional biographical informations and mathematical details extracted from the *Gesammelte Abhandlungen* [20] and from the fourth and fifth supplements of the *Grundzüge einer allgemeinen Theorie der linearen Integralgleichungen.* [13, 14] of David Hilbert. This emphasizes that we will not provide a complete overview of the lifelong friendship between Hilbert and Hurwitz or their extensive interdisciplinary exchange of mathematical ideas. However, since most of the documents stored in Zurich have not been processed yet, their perception should nevertheless extend the well known facts of their relationship. We will concentrate on the teacher–student aspect and in particular tackle the question of their mutual influence, following [35].

3.1 The Beginning of a Friendship

In 1884, when Hurwitz received his first full professorship in Königsberg, he and David Hilbert met for the first time. The younger wrote later: "His friendly and open nature won him, when he came to Königsberg, quickly the hearts of all who got to know him there [...]"[13] [17, p. 167]. Hilbert - born, grown up and studying in Königsberg[14] - was an extraordinarily inquisitive young mathematician, who craved for progressive mathematical knowledge. For him and his two-years younger friend Hermann Minkowski, who was said to be an exceptional talent, it was definitely a very fortunate coincidence that Adolf Hurwitz became their teacher. Since Hurwitz was not only familiar with the mathematical school created by Alfred Clebsch (1833–1872) and Felix Klein, but also had learned in Berlin from Leopold Kronecker and especially Karl Weierstrass (see Sect. 1), the contribution of his knowledge was enormous.[15] In fact, Hilbert himself mentioned in his obituary for Adolf Hurwitz [17, p. 162],

> Here I was, at that time still a student, soon asked for scientific exchange and had the luck by being together with him to get to know in the easiest and most interesting way the directions of thinking of the at time opposite however each other excellently complementing schools, the geometrical school of Klein and the algebraic-analytical school of Berlin. [...] On numerous, sometimes day by day undertaken walks at the time for eight years we have browsed through probably all corners of mathematical knowledge, and Hurwitz with his as well wide and multifaceted as also established and well-ordered knowledge was always our leader.[16]

Furthermore, in his colourful biography of David Hilbert, Otto Blumenthal[17] (1876–1944) quoted, "We, Minkowski and me, were quite overwhelmed with his knowledge and did not believe that we could ever get that far."[18] [3, p. 390]. However, Adolf

---

[13]"Sein freundliches und offenes Wesen gewann ihm, als er nach Königsberg kam, rasch die Herzen aller, die ihn dort kennenlernten [...]".

[14]With the exception of one year, 1881, at the University of Heidelberg (see [40]).

[15]In particular concerning both faces of complex analysis [3, p. 390].

[16]"Hier wurde ich, damals noch Student, bald von Hurwitz zu wissenschaftlichem Verkehr herangezogen und hatte das Glück, durch das Zusammensein mit ihm in der mühelosesten und interessantesten Art die Gedankenrichtungen der beiden sich damals gegenüberstehenden und doch einander so vortrefflich ergänzenden Schulen, der geometrischen Schule von Klein und der algebraisch-analytischen Berliner Schule kennenzulernen. [...] Auf zahllosen, zeitweise Tag für Tag unternommenen Spaziergängen haben wir damals während acht Jahren wohl alle Winkel mathematischen Wissens durchstöbert, und Hurwitz mit seinen ebenso ausgedehnten und vielseitigen wie festbegründeten und wohlgeordneten Kenntnissen war uns dabei immer der Führer."

[17]Hilbert's first doctoral student (in 1898).

[18]"Wir, Minkowski und ich, waren ganz erschlagen von seinem Wissen und glaubten nicht, dass wir es jemals so weit bringen würden."

Hurwitz was also concerned about his students. "In the lessons he always took great care by interesting exercises to motivate for participation, and it was characteristic, how often one could find him in his thoughts searching for appropriate exercises and problems."[19] [17, p. 166], remembered the former student Hilbert as well as, "Inspirations were given by the mathematical Colloquium [...] in particular, however, by the walks with Hurwitz "in the afternoon precisely at 5 o'clock next to the apple tree""[20] [3, p. 393]. This tradition of joint walks with students had been continued by Hilbert for all of his academic life. We may conclude that in the beginning it was naturally Hilbert who benefited a lot from his teacher Hurwitz. In 1892, he received his first professorship as successor of Hurwitz in Königsberg.

3.2 Two Productive Universal Mathematicians

Both mathematicians, Adolf Hurwitz as well as David Hilbert, belonged to a dying species in their profession: They can be considered as universal mathematicians having comprehensive knowledge and scientific results in various mathematical disciplines.[21] Furthermore, both were extremely productive. We can get an impression of their work by noticing that the *Gesammelten Abhandlungen I - III* of Hilbert [20] consist of more than 1350 pages of mostly influential mathematics similar to the *Mathematisches Werk I + II* of Adolf Hurwitz [27] having more than 1400 pages.

Without claiming to be exhaustive, we want to give a strongly shortened overview of the mathematical work of David Hilbert following the biographical essay *Lebensgeschichte* [3] written by his former doctoral student Otto Blumenthal. The subsequent list sketches Hilbert's wide scientific spectrum concerning all main mathematical disciplines, ordered by modern terms, highlighting some results, publications or speeches, which we want to analyze in respect to their connection to Hurwitz's work in the following subsection.

- 1885–1892 **Algebra**: theory of invariants
- 1890 *Ueber die Theorie der algebraischen Formen* [10]
- 1892 *Ueber die Irrationalität ganzer rationaler Funktionen mit ganzzahligen Koeffizienten* [20, vl. II, No. 18] with "Irreduzibilitätssatz" [3, p. 393]
- 1892–1899 **Number Theory**: theory of number fields
- 1893 Simplification of the Hermite-Lindemann proof of the transcendence of e and π [20, vl. I, No. 1]

---

[19]"In den Übungen war er ständig darauf bedacht, durch anregende Aufgaben zur Mitarbeit heranzuziehen, und es war charakteristisch, wie oft man ihn in seinen Gedanken auf der Suche nach geeigneten Aufgaben und Problemstellungen für Schüler antraf."

[20]"Anregungen vermittelten das Mathematische Kolloquium [...], vor allem aber die Spaziergänge mit Hurwitz "nachmittags präzise 5 Uhr nach dem Apfelbaum"".

[21]In view of the growth of the mathematical community and its insights around the turn to the twentieth century, Hilbert and Henri Poincaré are said to be the last knowing almost everything about the whole developments in mathematics.

- 1894 *Zwei neue Beweise für die Zerlegbarkeit der Zahlen eines Körpers in Primideale* [20, vl. I, No. 2]
- 1896 *Die Theorie der algebraischen Zahlkörper* [20, vl. I, No. 7], also called "Zahlbericht" including ideal theory
- 1891–1902 **Geometry**: axiomatization of geometry
- 1895–1903 *Grundlagen der Geometrie* including complements [18]
- 1895 *Über die gerade Linie als kürzeste Verbindung zweier Punkte* [18, compl. I]
- 1900 *Über den Zahlbegriff* [12]: axiomatization of arithmetic
- 1900 Hilbert stated his 23 mathematical problems at the International Congress of Mathematicians in Paris [11]
- 1902–1910 **Complex Analysis**: variation problems, independence theorem
- 1904–1910 **Linear Algebra, Functional Analysis**: *Grundzüge einer allgemeinen Theorie der linearen Integralgleichungen* with supplements [13, 14] including new terminology
- 1907 (published 1910) **Analysis meets Geometry**: analytical refounding of Minkowski's theory of volumes and surfaces of convex bodies in [16]
- 1907 **Analysis meets Number Theory**: *Beweis der Darstellbarkeit der ganzen Zahlen durch eine feste Anzahl nter Potenzen (Waringsches Problem)* [15]
- 1902–1918 Axiomatization of Physics and Mechanics: theory of relativity
- 1904–1934 **Mathematical Foundations**
- 1904 A first talk in Heidelberg about *Axiomatisierung der Zahlenlehre* [3, p. 421]
- 1922–1934 **Hilbert Program**: formalism and proof theory
- 1931 *Die Grundlegung der elementaren Zahlenlehre* [19]

Although David Hilbert and Adolf Hurwitz were very similar with respect to their extraordinary productivity, their different characters can be recognized in their academic behavior. Hilbert was noticed as extroverted and "became used to be a famous man"[22] [3, p. 407], whereas Hurwitz "avoided any personal being apparent in academic and public life"[23] [17, p. 167] and preferred to work continuously, however, silently. This is particularly visible in his consequent, nearly peerless way of taking notes of mathematical ideas in his diaries. In the following we take a glance at those diaries in view of parallels to the above listed fields of research of David Hilbert.

3.3 Mathematical Exchange

Already here, anticipating, it shall be reported about the seldom harmonic and fruitful cooperation of those three mathematicians.[24] [3, p. 390],

noted Otto Blumenthal and refered to the active mathematical exchange between Minkowski, Hilbert and Hurwitz. In the mathematical diaries [21] various entries,

---

[22]"gewöhnte sich daran, ein berühmter Mann zu sein".

[23]"Mied jedes persönliche Hervortreten im akademischen und öffentlichen Leben".

[24]"Es soll schon hier vorgreifend über das selten harmonische und fruchtbare Zusammenarbeiten dieser drei Mathematiker berichtet werden."

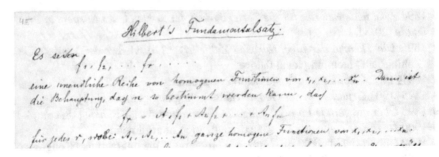

Fig. 8 "Hilbert's Fundamental Theorem. Let $f_1, f_2, \cdots, f_r, \ldots$ be an infinite series of homogeneous functions of $x_1, x_2, \cdots x_n$. We claim that n can be determined in such a way that $f_r = A_1 f_1 + A_2 f_2 + \cdots A_n f_n$ for any r, where $A_1, A_2, \cdots, A_n$ are entire homogeneous functions of $x_1, x_2, \cdots, x_n$." ("Hilbert's Fundamentalsatz. Es seien $f_1, f_2, \ldots, f_r, \ldots$ eine unendliche Reihe von homogenen Funktionen von $x_1, x_2, \ldots x_n$. Dann ist die Behauptung, daß n so bestimmt werden kann, daß $f_r = A_1 f_1 + A_2 f_2 + \cdots A_n f_n$ für jedes r, wobei $A_1, A_2, \ldots, A_n$ ganze homogene Functionen von $x_1, x_2, \ldots, x_n$.")

directly or indirectly related to publications of Hilbert, can be found - those are listed in Appendix "Appendix I: Links to Hilbert in the ETH Estate of Hurwitz". Furthermore, some ideas in selected diaries suggest to be inspired by Hilbert.

We already pointed out that without any doubt in their first years it was essentially David Hilbert who benefited from his teacher. However, his teacher Hurwitz became very soon aware of his talented student. In diary No. 6^{25} [21, No. 6] on page 44 is a first entry related to Hilbert, entitled "On Noether's Theorem (concerning a message of Hilbert)".[26] Here Hurwitz familiarized himself with the nowadays called residual intersection theorem, sometimes also fundamental theorem, of Max Noether (1844–1921), dealing with a linear form associated with two algebraic curves. Interestingly one page later follows the entry "Hilbert's Fundamental Theorem"[27] dealing with a linear form of homogeneous functions (Fig. 8).

On the one hand, this can be considered as continuation and extension of Noether's theorem, on the other hand, this is a previous version of Hilbert's basis theorem. Although Hurwitz was obviously interested perhaps even inspired from Hilbert's form and invariant theory, in his work *Über die Erzeugung der Invarianten durch Integration*.[28] Hurwitz discovered a "new generating principle for algebraic invariants which allowed him to apply an [by Hilbert] introduced method [...]."[29] [17, p. 164]. This nice formulation comes from Hilbert himself and can be interpreted in such a way that Hurwitz kind of improved the use of one of Hilbert's methods. At least three more entries in the mathematical diaries (see Appendix II: No. 8, p. 207;

---

[25]From 1888 IV. to 1889 XI.

[26]"Der Nöther'sche Satz (nach einer Mitteilung von Hilbert)".

[27]"Hilberts Fundamentalsatz".

[28]"On generating invariants by integration."

[29]"Neues Erzeugungsprinzip für algebraische Invarianten, das es ihm ermöglicht, ein [von Hilbert] eingeschlagenes Verfahren [...] anzuwenden."

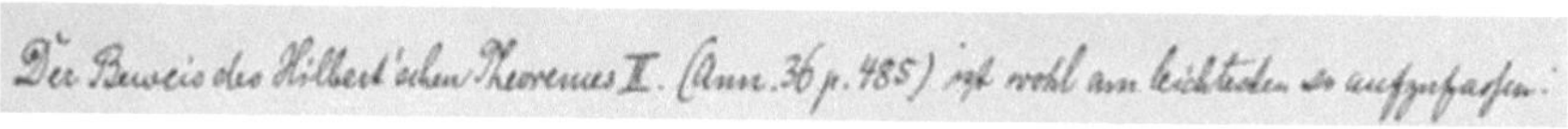
Der Beweis des Hilbert'schen Theorems II. (Ann. 36 p. 485) ist wohl am leichtesten so aufzufassen:

Fig. 9 "The proof of Hilbert's theorem (Ann 36. p. 485) seems to be the most easiest understandable in such a way: [...]" ("Der Beweis des Hilbert'schen Theorems II (Ann 36. p. 485) ist wohl am leichtesten so aufzufassen: [...]")

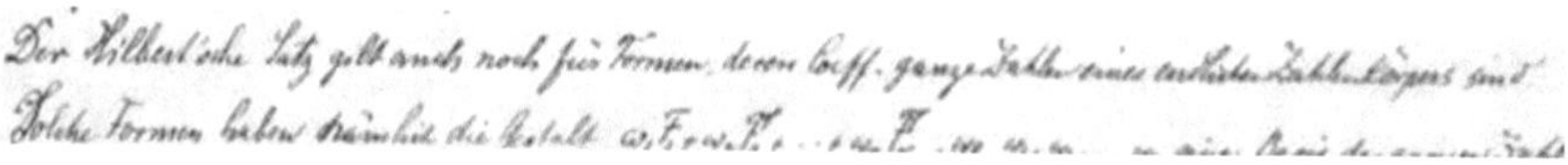
Der Hilbert'sche Satz gilt auch noch für Formen, deren Coeff. ganze Zahlen eines endlichen Zahlkörpers sind.
Solche Formen haben nämlich die Gestalt [illegible]

Fig. 10 "Hilbert's theorem holds also for forms whose coeff. are integers of a finite number field." ("Der Hilbert'sche Satz gilt auch noch für Formen, deren Coeff. ganze Zahlen eines endlichen Zahlkörpers sind.")

No. 14, p. 204; No. 25, p. 77) are directly dedicated to Hilbert's "Formensatz", nowadays called basis theorem. In the entry of the fourteenth diary[30] [21, No. 14], Hurwitz's comments on Hilbert's theorems have the tendency to sound like amendments. In the beginning, he wrote (Fig. 9)

Here Hurwitz sounds as if Hilbert's ideas respectively the way Hilbert had put the proof to language could be simplified. Another example is on the next page (Fig. 10):

This generalization of the theorem of Hilbert is remarkable, because it obviously paved the way for one of the later known versions of Hilbert's basis theorem: *The ring of polynomials* $K[X_1, \ldots, X_n]$ *over a field* K *is Noetherian.*[31]

In 1891, Hurwitz and Hilbert published a first and last joint note [20, vl. II, Nr. 17] in which they observed "that a certain, a number of parameters including irreducible ternary form still is irreducible for general integral values of those parameters."[32] [3, p. 393] Otto Blumenthal considered this as foundation of Hilbert's famous "Irreduzibilitätssatz" [20, vl. II, Nr. 18] from 1892.

In his first years being *Privatdozent*[33] in Königsberg, when Hilbert adressed himself to the theory of number fields, he reported from his and Adolf Hurwitz's common walks, discussing theories of Dedekind and Kronecker. "One considered Kronecker's proof for a unique decomposition of prime ideals, the other the one of Dedekind, and we thought both were awful."[34] [3, p. 397]. However, their cooperation turned out to be successfull: Firstly, Hilbert began a publication building on his talk *Zwei*

---

[30]From 1896 I.1. to 1897 II.1.

[31]We may assume that Hurwitz had meant "algebraic" instead of "finite".

[32]"daß eine gewisse, eine Anzahl Parameter enthaltende irreduzible ternäre Form auch für allgemeine ganzzahlige Werte dieser Parameter irreduzibel bleibt."

[33]In the German system habilitation granted the "venia legendi", i.e., the permission to lecture as *Privatdozent* which at that time meant to collectcourse fees from the students without any payment from the university.

[34]"Einer nahm den Kroneckersches Beweis für die eindeutige Zerlegung in Primideale vor, der andere den Dedekindschen, und beide fanden wir scheußlich."

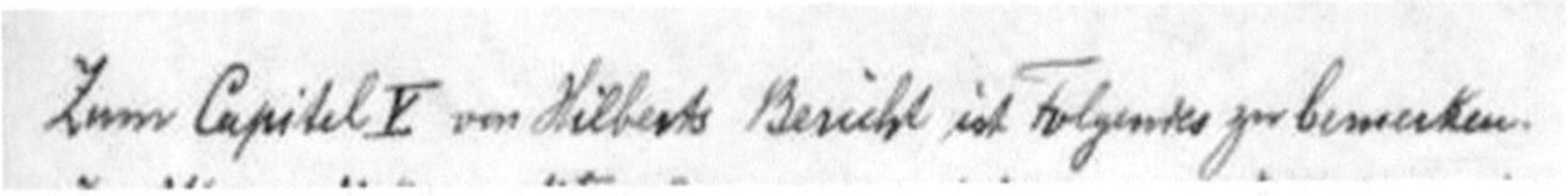

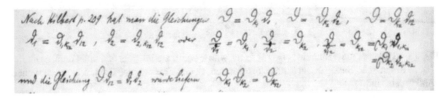

Fig. 11 "Concerning Chapter V of Hilbert's report the following is to remark." ("Zum Capitel V von Hilberts Bericht ist Folgendes zu bemerken.")

Fig. 12 "According to Hilbert p. 209 we have [...] and the equation $\nu\nu_{12} = \nu_1\nu_2$ would lead to $\nu_{k_1}\nu_{k_2} = \nu_{k_{12}}$" ("Nach Hilbert p. 209 hat man [...] und die Gleichung $\nu\nu_{12} = \nu_1\nu_2$ würde liefern $\nu_{k_1}\nu_{k_2} = \nu_{k_{12}}$")

neue Beweise für die Zerlegbarkeit der Zahlen eines Körpers in Primideale.[35] Secondly, Hurwitz, also working on algebraic number fields, published another proof later in his paper *Der Euklidische Divisionssatz in einem endlichen algebraischen Zahlkörper.*[36] Hilbert considered this work as "remarkable in view of the analogy with the euclidean algorithm in number theory"[37] [17, p. 165] and even preferred Hurwitz's proof in his famous *Zahlbericht.*[38] It seems that even after Hurwitz's moving to Zurich respectively Hilbert's full professorship in Königsberg starting from 1892, Hilbert was still slightly influenced by his former teacher. We investigate this also on behalf of two entries from 1898 and 1899 in Hurwitz's diaries No. 15[39] [21, No. 15] and No. 16[40] [21, No. 16] directly refering to Hilbert's *Zahlbericht* (Fig. 11). The first is listed as "Concerning Hilbert's "Report on Number Fields"".[41]

The subsequent entry gives the impression that Hurwitz continued Hilbert's work using a result of Hilbert about composing ideals of subfields of number fields. Hereafter, number fields will be defined by K respectively k_i and ideals by ν respectively ν_i, $i = 1, 2, 12$ (Fig. 12).

Then Hurwitz first verified this consequence $\nu_{k_1}\nu_{k_2} = \nu_{k_{12}}$ of Hilbert's formula, before he compared it with his conjecture (Fig. 13).

---

[35] *"Two new proofs of the decomposability of numbers of a number field in prime ideals"*, given in September 1893 at the meeting of the "Deutsche Mathematiker Vereinigung" in Munich.

[36] *"The Euclidean division theorem in a finite algebraic number field."*.

[37] "bemerkenswert durch die Analogie mit dem Euklidischen Algorithmus in der elementaren Zahlentheorie".

[38] Actually *Die Theorie der algebraischen Zahlkörper*, report of algebraic number theory.

[39] From 1897 II.1. to 1898 III.19.

[40] From 1898 III.20. to 1899 II.23.

[41] "Zu Hilbert's "Körperbericht"".

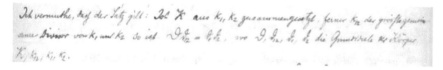

Fig. 13 "I assume that the theorem holds: Is K a composition of k_1, k_2, moreover k_{12} the greatest common divisor of k_1 and k_2 then $\nu\nu_{k_{12}} = \nu_{k_1}\nu_{k_2}$, where $\nu, \nu_{12}, \nu_1, \nu_2$ are Grundideale of the number fields K, k_{12}, k_1, k_2." (The use of the notation "Grundideal" here is a bit confusing. According to [41] it goes back to a work of Emmy Noether from 1923. However, this comment is misleading. According to [20, vl. I, p. 90] the term "Grundideal" was already used by Dedekind and Hilbert himself invented the new term "Differente" which is still used today. Interestingly, Hurwitz refering to Hilbert kept Dedekind's notation. "Ich vermute, daß der Satz gilt: Ist K aus k_1, k_2 zusammengesetzt, ferner k_{12} der größte gemeinsame Divisor von k_1 und k_2 so ist $\nu\nu_{k_{12}} = \nu_{k_1}\nu_{k_2}$, wo $\nu, \nu_{12}, \nu_1, \nu_2$ die Grundideale der Körper K, k_{12}, k_1, k_2.")

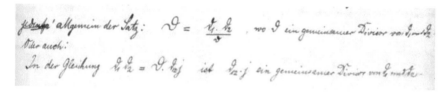

Fig. 14 "[...] consequently the general theorem holds: $\nu = \frac{\nu_1 \cdot \nu_2}{\nu}$, with ν a common divisor of ν_1 und ν_2. Or also: in the equation $\nu_1\nu_2 = \nu\nu_{12} \cdot j$ is $\nu_{12} \cdot j$ a common divisor of ν_1 and ν_2." (probably ν_{12} and ν were interchanged "[...] folglich besteht allgemein der Satz: $\nu = \frac{\nu_1 \cdot \nu_2}{\nu}$, wo ν ein gemeinsamer Divisor von ν_1 und ν_2. Oder auch: In der Gleichung $\nu_1\nu_2 = \nu\nu_{12} \cdot j$ ist $\nu_{12} \cdot j$ ein gemeinsamer Divisor von ν_1 und ν_2.")

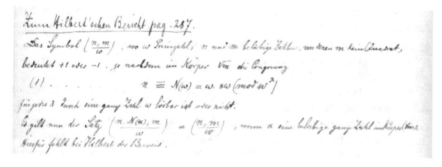

Fig. 15 Under the heading "Concerning Hilbert's report pag. 287" the symbol $\left(\frac{n,m}{\omega}\right)$, where ω is a prime number, n and m are arbitrary numbers and m is not a square number is treated. ("Zum Hilbert'schen Bericht pag. 287")

Consequently, Hurwitz deduced a general new result on algebraic number fields (Fig. 14).

Here Hurwitz used the *Zahlbericht* as a textbook, and the entry in No. 16 [21, No. 16] emphasizes that the teacher Hurwitz reflects on Hilbert's report (Fig. 15).

It is defined to be equal to $+1$ or -1, according to the property whether in the field $\mathbb{Q}(\sqrt{m})$ the congruence

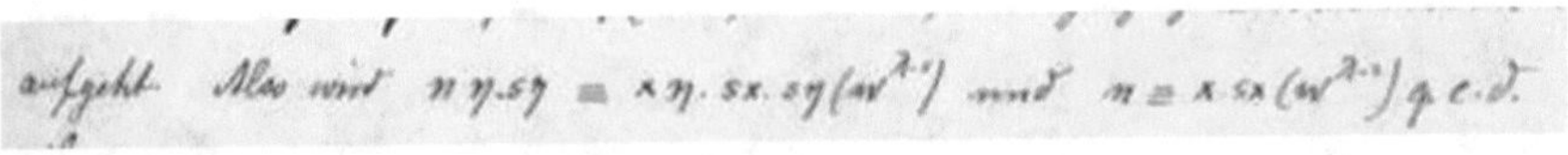

Fig. 16 Excerpt of Hilbert's *Zahlbericht*, page 289 respectively [20, vl. I, p. 164]

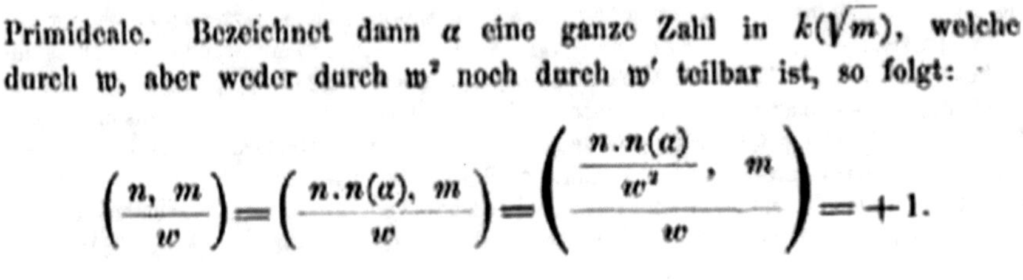

Primideale. Bezeichnet dann α eine ganze Zahl in $k(\sqrt{m})$, welche durch $\mathfrak{w}$, aber weder durch $\mathfrak{w}^2$ noch durch $\mathfrak{w}'$ teilbar ist, so folgt:

$$\left(\frac{n,\ m}{\mathfrak{w}}\right) = \left(\frac{n.n(\alpha),\ m}{\mathfrak{w}}\right) = \left(\frac{\dfrac{n.n(\alpha)}{\mathfrak{w}^2},\ m}{\mathfrak{w}}\right) = +1.$$

Fig. 17 "Thus $nn \cdot sn \equiv xn \cdot sx \cdot sn(\omega^{\lambda-1})$ and $n \equiv x \cdot sx(\omega^{\lambda-2})$ q.e.d." ("Also wird $nn \cdot sn \equiv xn \cdot sx \cdot sn(\omega^{\lambda-1})$ und $n \equiv x \cdot sx(\omega^{\lambda-2})$ q.e.d.")

$$n \equiv N(\omega) = \omega \cdot s\omega (\mathrm{mod}\, \omega^\lambda)$$

has for any λ a solution for an integer number ω or not (here s and λ are not defined precisely).[42] Hurwitz stated, "So we have the theorem

$$\left(\frac{n \cdot N(\alpha), m}{\omega}\right) = \left(\frac{n, m}{\omega}\right),$$

if α is an arbitrary entire number in the number field $(\sqrt{m})$."[43,44] and continued "Therefore Hilbert lacks a proof."[45] Indeed, we can find the stated equation some pages later in Hilbert's work without a sound verification (Fig. 16):

Within one page Hurwitz filled Hilbert's gap proving this equation by use of a clever case distinction (Fig. 17).

Obviously, Hurwitz worked not only with Hilbert's *Zahlbericht*, moreover, he obtained some improvements.

Another diary entry has to be mentioned for the sake of completeness, however any conclusion would be rather speculative: In No. 23 [21, No. 23] from 1908 Hurwitz dealt with a short exercise entitled *Über die kürzeste Linie auf einem Parallelepiped*. Remarkably one of Hilbert's papers from 1895 has the very similar title *Über die*

---

[42] Hilbert characterized with his symbol so-called "Normenreste" respectively "Normennichtreste" [20, vl. I, p. 164] of a number field. Today the symbol is known as "Hilbert symbol".

[43] "Es gilt nun der Satz

$$\left(\frac{n \cdot N(\alpha), m}{\omega}\right) = \left(\frac{n, m}{\omega}\right),$$

wenn α eine beliebige ganze Zahl im Körper $(\sqrt{m})$."

[44] We may assume that an algebraic number field $K(\sqrt{m})$ was meant here.

[45] "Hierfür fehlt Hilbert der Beweis."

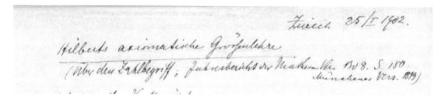

Fig. 18 The entry is entitled "Hilbert's axiomatic theory of quantities" ("Hilbert's axiomatische Größenlehre"), where Hurwitz refered to Hilbert's talk *Über den Zahlbegriff* [12]

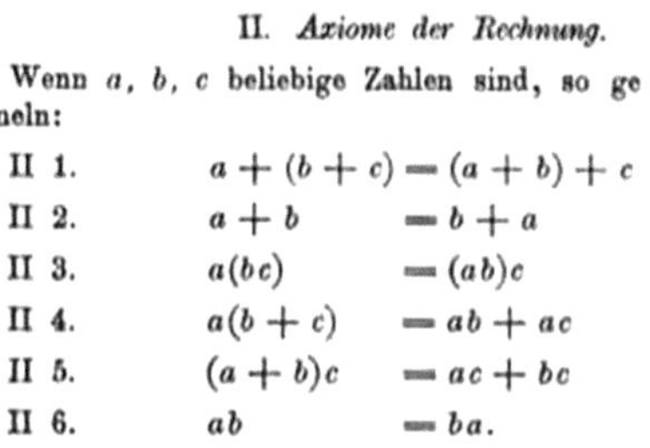

II. *Axiome der Rechnung.*

Wenn a, b, c beliebige Zahlen sind, so ge-meln:

II 1.	$a + (b + c) = (a + b) + c$
II 2.	$a + b \quad = b + a$
II 3.	$a(bc) \quad = (ab)c$
II 4.	$a(b + c) \quad = ab + ac$
II 5.	$(a + b)c \quad = ac + bc$
II 6.	$ab \quad = ba.$

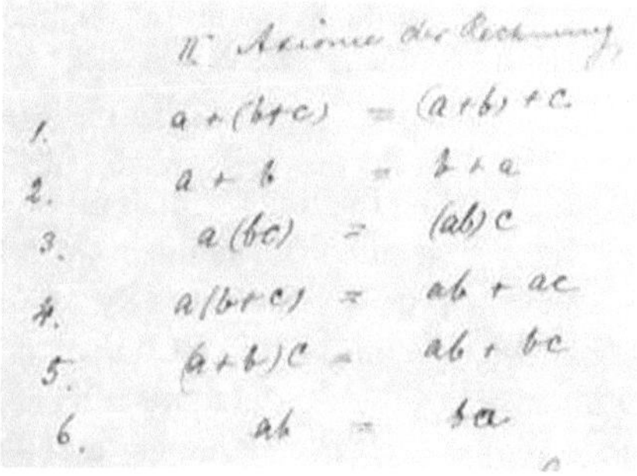

Fig. 19 Excerpts of Hilbert's publication [12] and Hurwitz's diary entry, "II. Axioms of Calculation. [...]"

gerade Linie als kürzeste Verbindung zweier Punkte. However, since these two topics were examined with a distance of 13 years, we will not take this coincidence into consideration (Fig. 18).

Instead, we go on chronologically and take a look at diary No. 19[46] [21, No. 19] from 1902.

After his axiomatization of geometry, Hilbert continued working on developing an axiomatic system for arithmetic. He was demanding for complete freedom from contradictions in mathematics. Therewith, Hilbert became one of the founders of a new philosophical movement in mathematics, the complete establishment of mathematics on an axiomatic system, being the first advocate of the so-called formalism.[47] His above mentioned talk and subsequent publication [12] can be considered as milestone. What makes Hurwitz's entry so interesting, is that it contains nothing but a nearly exact copy of Hilbert's ideas (Fig. 19).

Here Hurwitz followed step by step, axiom by axiom, Hilbert's rules for operations, calculations, order and continuity and even his consequences, as well as Hilbert's new terminology (Figs. 20 and 21),

It seems that this concept was completely new for Hurwitz and, furthermore, that he was willing to understand Hilbert's axiomatization. Here we get a first idea that

---

[46]From 1901 XI.1. to 1904 III.16.

[47]Due to inspirations of various mathematicians his attitude was in a steadily evolvement. For more information we refer to [47].

Fig. 20 "Some remarks on the dependence of axioms were added by Hilbert:" ("Einige Bemerkungen über die Abhängigkeit der Axiome hat Hilbert hinzugefügt:")

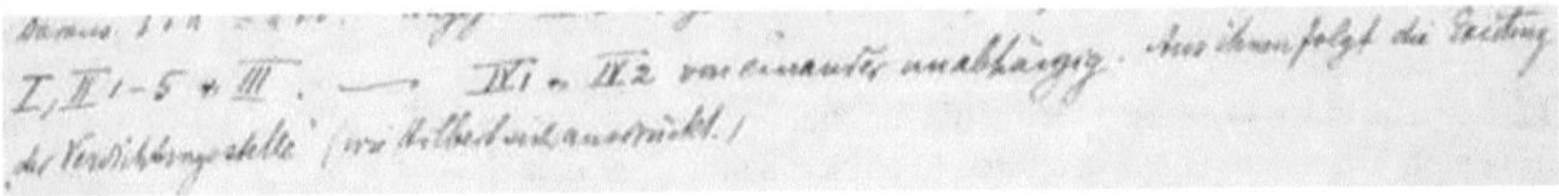

Fig. 21 "From them the existence of a "Verdichtungsstelle" follows (as Hilbert expresses himself.)" ("Aus ihnen folgt die Existenz der "Verdichtungsstelle" (wie Hilbert sich ausdrückt.)")

the status of their relation had become collegial. In any case, David Hilbert and Adolf Hurwitz are on an equal footing at the turn of the century.

The year 1900 was also the year when Hilbert gave his talk about his famous 23 problems at the International Conference of Mathematicians in Paris [11]. It might be interesting to notice that one entry in diary No. 22 [21, No. 22] concerning the proof of the Theorem of Pythagoras could have been inspired by the third problem.

More obvious, however, is Hilbert's influence on an entry in diary No. 21[48] [21, No. 21] from August 09, 1906, which is entitled "D. Hilbert (Integralequ. V. Gött, Nachr. 1906)"[49] and refers to the fifth supplement of Hilbert's article *Grundzüge einer allgemeinen Theorie der linearen Integralgleichungen* [14] from 1906. In this work Hilbert defined a new terminology and presented an innovative concept of handling linear algebra problems by applying integral equations. His method is based on the symmetry of the coefficients which is equivalent to the symmetry of the kernel[50] of the integral equation (see [3, p. 411]). The fourth [13] and fifth supplement, to which Hurwitz refered, extend Hilbert's previous results on bilinear forms with infinitely many variables. The diary entry begins with the section "Bezeichungen", which means "notations". One after the other Hurwitz reproduced Hilbert's definitions of the terms "Abschnitte" and "Faltung" as well as "Eigenwerte", "Spektrum" and "Resolvente" (Fig. 22).[51]

In fact, those terms were introduced by Hilbert himself. On page 459 of the fifth supplement he wrote: "Values are significantly determined by the kernel(s,t); I named them *Eigenwerte* resp. *Eigenfunktionen* [...]."[52] It seems that Hurwitz was not yet familiar with Hilbert's concept. One indication is the special highlighting of all those terms by double underlining, another indication is the placing of certain

---

[48] From 1906 II.1. to 1906 XII.8.

[49] "D. Hilbert (Integralgl. V. Gött. Nachr. 1906)".

[50] Hilbert's term "Kern" became internationally used, in English it was transformed to "kernel".

[51] "Segments", "convolution", "eigenvalue", "spectrum", "resolvent".

[52] "Die Werte [...] sind wesentlich durch den Kern (s,t) bestimmt; ich habe sie *Eigenwerte* bez. *Eigenfunktionen* [...] genannt." In English these objects are now called *eigenvalues* and *eigenfunctions*.

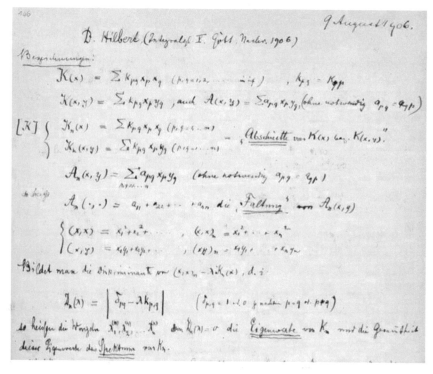

Fig. 22 Excerpt of Hurwitz's diary entry. The important terms are double underlined

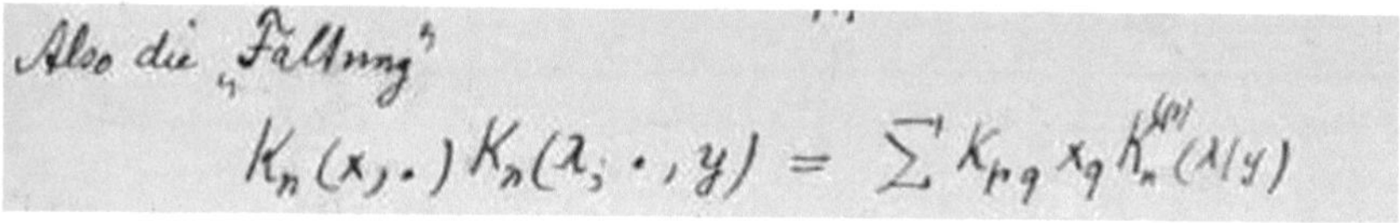

Fig. 23 "Thus the "convolution" $K_n(x, \cdot)K_p(\lambda_j\cdot, y) = \sum K_{pq}x_q K_n^{(p)}(\lambda/y)$" ("Also die "Faltung" $K_n(x, \cdot)K_p(\lambda_j\cdot, y) = \sum K_{pq}x_q K_n^{(p)}(\lambda/y)$")

terms in quotation marks on the second page, where Hurwitz became acquainted with resulting equations (Fig. 23).

This shows the unfamiliar use of this term. Some days later, in an entry from September 02, 1906, Hurwitz continued his analysis. He wrote,

Then Hurwitz defined again a variety of new terms. This is followed up through an entry from September 16, 1906, where Hurwitz stated a theorem on quadratic forms (Figs. 24 and 25).

Hurwitz examined the theorem as well as the proof and found some reformulations, noted on diary pages 170 and 171. Two pages later he stated a further theorem (Fig. 26).

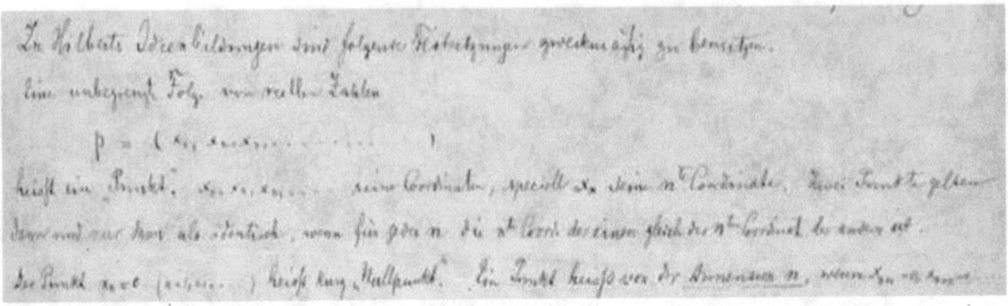

Fig. 24 "In Hilbert's concept the following continuations are expediently to be used [...]" ("In Hilberts Ideenbildung sind folgende Fortsetzungen zweckmäßig zu benutzen [...]")

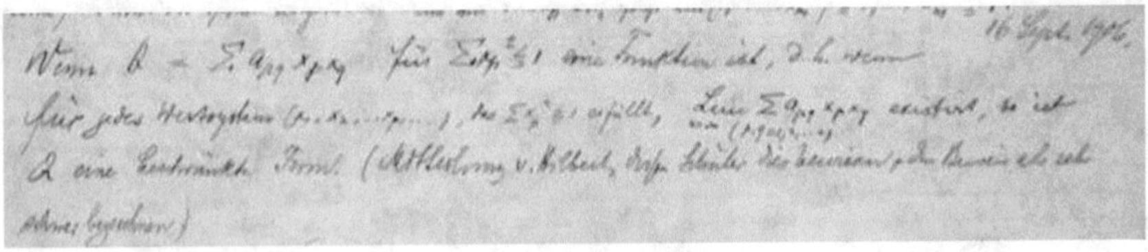

Fig. 25 "If $Q = \sum a_{pq}x_px_q$ for $\sum x_p^2 \leq 1$ is a function, i.e. if for any system of values $(x_1, x_2, \cdots, x_p, \cdots)$, which satisfies $\sum x_p^2 \leq 1$, $Lim_{n=\infty(p,q=1,2,\ldots n)} \sum a_{pq}x_px_q$ exists, then Q is a limited form. (notification of Hilbert, that students proved this, the proof is to be considered as very difficult.)" ("Wenn $Q = \sum a_{pq}x_px_q$ für $\sum x_p^2 \leq 1$ eine Funktion ist, d.h. wenn für jedes Wertsystem $(x_1, x_2, \cdots, x_p, \cdots)$, das $\sum x_p^2 \leq 1$ erfüllt, $Lim_{n=\infty(p,q=1,2,\ldots n)} \sum a_{pq}x_px_q$ existiert, so ist Q eine beschränkte Form. (Mitteilung v. Hilbert, daß Schüler dies bewiesen, den Beweis als sehr schwer bezeichnen.)")

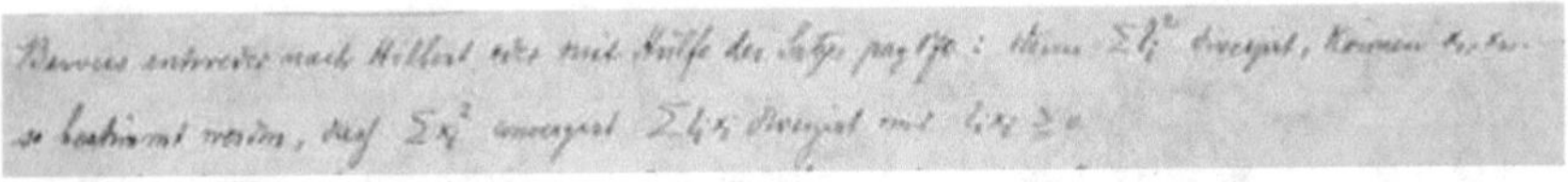

Fig. 26 "Proof either according to Hilbert or with the theorem on pag 170 [...]" ("Beweis entweder nach Hilbert oder mit Hilfe des Satzes pag 170 [...]".)

The next section continues with the consideration of linear and quadratic forms, which we do not want to deepen here. However, what we remark is that Hurwitz was not only dealing with Hilbert's ideas, he was moreover adopting a completely new theory from his colleague.

One year later the collegial relation became more apparent. In 1907, David Hilbert solved the task of developing an analytical refounding of Minkowski's theory of volumes and surfaces of convex bodies in his sixth supplement[53] [16]. This task had already been tackled by Hurwitz in 1901 and 1902. An entry in diary No. 18[54] [21, No. 18] about his colloquium talk from January 21, 1901 is entitled (Fig. 27).

---

[53]Which was published three years later in 1910.
[54]From 1900 XII. to 1901 X.

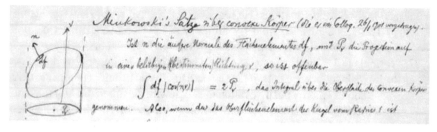

Fig. 27 "Minkowski's theorems on convex bodies [...]". ("Minkowski's Sätze über konvexe Körper [...]")

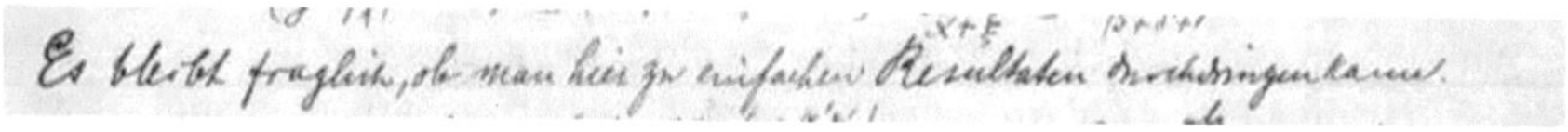

Fig. 28 "It remains doubtful if simple results can be discovered here." ("Es bleibt fraglich, ob man hier zu einfachen Resultaten durchdringen kann.")

Here Hurwitz discussed several questions on convex bodies. However, four pages of calculations later, he stated (Fig. 28).

Obviously, he was not content with his considerations. One year later, Adolf Hurwitz published the article [25] in which he tried a first attempt of an appropriate refoundation "using his theory of spherical functions [...], however, he only had a partial success. Hilbert, with his powerful tool on integral equations, replaces the spherical function by more generalized ones and passes through."[55] [3, p. 414].[56]

Hilbert developed more and more as guidepost and unique mathematician. At the latest, the year 1907 can be considered as final turnaround of the teacher–student relation to Hurwitz. A therefore remarkable situaton is explained in detail by Otto Blumenthal: in a short note [26] from November 20, 1907, Adolf Hurwitz proved a variation of the so-called Waring problem. This number theoretical question, named after the English mathematician Edward Waring (1736–1798) and published in his work *Meditationes algebraicae* from 1770, claimed that for every exponent $n \in \mathbb{N}$ there exists a natural number m such that every natural number can be expressed as a sum of at least m many n-th powers. Hurwitz showed, "[i]s the nth power of $x_1^2 + x_2^2 + x_3^2 + x_4^2$ equal to a sum of $2n$th powers of a linear rational form of x_1, x_2, x_3, x_4, and does the Waring Conjecture hold for n, it is also valid for $2n$."[57] [3, p. 415] According to Otto Blumenthal, "[t]his theorem gave Hilbert the inspiration

---

[55]"mit seiner Theorie der Kugelfunktionen [...], hatte aber nur einen Teilerfolg erzielt. Hilbert, im Besitze der mächtigen Hilfsmittel der Integralgleichungen, ersetzt die Kugelfunktionen durch allgemeinere, und kommt durch."

[56]Firstly, it is remarkable that Hurwitz, studying Hilbert's supplements, did not apply the integral equation method. Secondly, notice that Hilbert's dissertation thesis was about spherical functions. Interestingly, the thesis is dedicated to Hurwitz.

[57]"Ist die n-te Potenz von $x_1^2 + x_2^2 + x_3^2 + x_4^2$ identisch gleich einer Summe $(2n)$-ter Potenzen linearer rationaler Formen der x_1, x_2, x_3, x_4, und gilt die Waringsche Behauptung für n, so gilt sie auch für $2n$."

and direction for his examinations. He found an unexpected way to state an identity of the by Hurwitz demanded kind for arbitrary n."[58] Moreover, Hilbert deduced "from a general principle which was used by Hurwitz in the theory of invariants in 1897, a formula [...]"[59] and he managed "to transfer the by the integration demanded taking the limits in the coefficients of the sum and finally, on behalf of another trick, to replace those coefficients by positive rationals. Therewith the foundation for the proof of Waring's theorem is laid."[60] [3, p. 415] Finally, Hilbert solved Waring's Problem [15] and presented a wonderful example for his mathematical skills: "[b]ecause he fought together with a master of Hurwitz's high level and won with the weapons from Hurwitz's armor chamber on a point, when [Hurwitz] had no prospect of success."[61] [3, p. 416]

With this meaningful characterization we close the analysis of the diary entries related to the mathematical exchange between those two great mathematicians with some last significant words from Hilbert, expressing that Hurwitz was "[...] more than willing to appreciate the achievements of others and he was genuinely pleased about any scientific progress: an idealist in the good old-fashioned meaning of the word."[62] [17, p. 164]

3.3.1 Personal Relation: Lifelong and Even Longer

Besides the mathematical diaries there are some more hints on the multifaceted relationship between David Hilbert and Adolf Hurwitz to be discovered in the ETH estate in Zurich. Hurwitz, who suffered during all of his life from an unstable health, was a very rare guest at conferences outside Zurich. Accordingly several greeting cards[63] sent from mathematical events can be found, from "Lutetia Parisiorum, le 12 août 1900", the "Landau-Kommers 18. Jan. 1913" and from the "Dirichletkommers am 13. Februar 1905".

All of those were signed by a great number of mathematicians, the first two show the handwriting of David Hilbert. On the card from Paris (see the left picture in

---

[58]"Dieser Satz gab Hilbert die Anregung und Richtung zu seinen Untersuchungen. Er fand nämlich einen ungeahnten Weg, um für beliebige n eine Identität der von Hurwitz geforderten Art aufzustellen."

[59]"aus einem allgemeinen Prinzip, das Hurwitz 1897 in der Invariantentheorie benutzt hatte, eine Formel [...]".

[60]"den durch die Integration geforderten Grenzübergang in die Koeffizienten der Summe zu verlegen und schließlich durch einen weiteren Kunstgriff diese Koeffizienten durch positive rationale zu ersetzen. Damit ist die Grundlage für den Beweis des Waringschen Satzes gelegt."

[61]"Denn er kämpfte zusammen mit einem Meister von dem hohem Range Hurwitz's und siegte mit den Waffen aus Hurwitz's Rüstkammer an einem Punkte, wo dieser keine Aussicht auf Erfolg gesehen hatte."

[62]"[...] gern bereit zur Anerkennung der Leistungen anderer und von aufrichtiger Freude erfüllt über jeden wissenschaftlichen Fortschritt an sich: ein Idealist im guten altmodischen Sinne des Wortes."

[63]Under the directory HS 583: 52,53 and 57.

Fig. 29   Greeting cards from the "Lutetia Parisiorum" and the "Landau-Kommers"

Fig. 29) he wrote, "Sending warm greetings, wishing good recovery and hoping for a soon reunion longer than the last time Hilbert".[64]

Furthermore, some documents testify that there had even been a close relationship between the families Hurwitz and Hilbert. As already mentioned above, Adolf Hurwitz's elder brother Julius, who also studied mathematics in Königsberg, edited several of Hilbert's lectures. On some lecture notes[65] comments of Hilbert himself can be found and in the extensive collected correspondence of Adolf Hurwitz[66] a letter exchange between Julius Hurwitz and Hilbert has been discovered.

In an additional notebook No. 32[67] Georg Pólya completed the mathematical diaries with a register. Next to this list the notation "the first 9 volumes and table of contents are for the purpose of editing temporarily at Prof. Hilbert in Göttingen"[68] and is written and later crossed out with a pencil. Obviously, Hilbert had lent the first nine as well as the twentysecond diary and had returned them after a while. It is difficult to reconstruct when exactly he borrowed the diaries, however, there are two hints. In the beginning of this section we already stated Hilbert's remarks about Hurwitz's diaries. He wrote that they "provide a true view of his constantly progressive development and at the same time they are a rich treasure trove for interesting and for further examination appropriate thoughts and problems." [17, p. 166] Since this quotation is taken from his commemorative speech, Hilbert had viewed the diaries before 1920 and considered them as a rich source of new mathematical inspirations (Fig. 30).

In a letter of condolence to Ida Samuel-Hurwitz[69] from December 15, 1919 - four weeks after Adolf Hurwitz's death - Hilbert wrote that for George Pólya (1887–1985)

---

[64]"Herzliche Grüße sendend, gute Erholung wünschend und baldiges Wiedersehen auf länger, wie das letzte Mal erhoffend. Hilbert".

[65]Under the directory HS 582: 154.

[66]Which is stored in the archive of Göttingen.

[67]In HS 582: 32.

[68]"die ersten 9 Bände und Inhaltsverzeichnis sind zwecks Bearbeitung vorderhand bei Prof. Hilbert in Göttingen".

[69]Under the directory HS 583: 28.

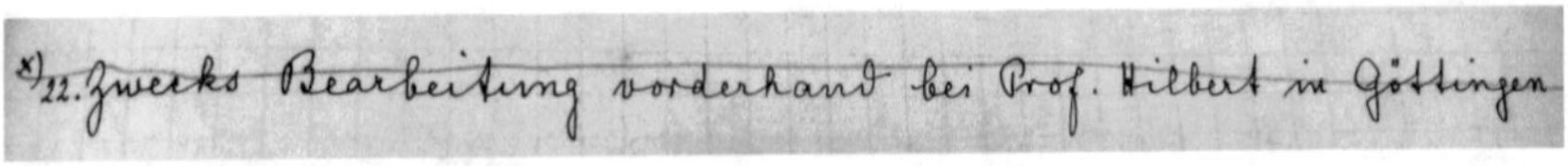

Fig. 30 "22. for editing temporarily at Prof. Hilbert in Göttingen" ("22. zwecks Bearbeitung vorder-hand bei Prof. Hilbert in Göttingen")

and him the "matter of publishing the Hurwitz's treatises [is] of utmost concern".[70,71] He offered, "The negotiations could be done verbally with Springer by a local, very skillful, math. colleague."[72,73] It took another few years before Hilbert's and Pólya's support of their mathematical and personal friend finally turned out to be successfull. Adolf Hurwitz's *Mathematische Werke* [27] were published in 1932.

4 Julius Hurwitz and an Ergodic Theoretical View on His Complex Continued Fraction

Reading Shigeru Tanaka's article *A complex continued fraction transformation and its ergodic properties* [46] with its modern terminology, one might get the impression that his algorithm provides a typical recent, innovative approach to complex continued fractions. However, it turns out that Tanaka's continued fraction expansion equals the one of Julius Hurwitz [30]. In this section we adopt the modern perspective to receive new tools for handling some characteristics concerning its approximation quality.

4.1 Basics

Tanaka's approach relies on the following representation of complex numbers z,

$$z = x\alpha + y\overline{\alpha},$$

with $x, y \in \mathbb{R}$ and $\alpha = 1 + i$, respectively $\overline{\alpha} = 1 - i$. Obviously, all complex numbers z have therewith a representation

$$z = x(1 + i) + y(1 - i) = (x + y) + i(x - y), \tag{4.1}$$

---

[70]"Angelegenheit der Herausgabe der Abhandlungen von Hurwitz [ist] unsere wichtigste Sorge".

[71]Pólya himself remembered, "I played a large role in editing his collected works." [39, p. 25].

[72]"Die Verhandlungen könnte ich durch einen hiesigen sehr gewandten math. Kollegen mündlich mit Springer führen lassen."

[73]Probably, Hilbert meant Richard Courant, his former student and at that time professor in Göttingen, with whom he had created the "Gelbe Buchreihe" with publisher Springer.

where x and y are uniquely determined by solving the system of linear equations $\operatorname{Re} z = x + y$ and $\operatorname{Im} z = x - y$. To facilitate the following calculations, we illustrate this representation of complex numbers. We have

$$z = a + ib = \left(\frac{1}{2}(a+b)\right)\alpha + \left(\frac{1}{2}(a-b)\right)\overline{\alpha},$$

which leads to

$$\begin{pmatrix} x \\ y \end{pmatrix} = \frac{1}{2}\begin{pmatrix} 1 & 1 \\ 1 & -1 \end{pmatrix}\begin{pmatrix} a \\ b \end{pmatrix}.$$

Accordingly, the sets

$$I_0 := \{n\alpha + m\overline{\alpha} \ : \ n, m \in \mathbb{Z}\},$$

respectively

$$I := I_0 \setminus \{0\}$$

describe subsets of the set of Gaussian integers $\mathbb{Z}[i]$. Since

$$n + m \equiv n - m \bmod 2 \tag{4.2}$$

for any pair of integers n, m, it follows that I_0 contains exactly those numbers z whose real part and imaginary part have even distance. Writing I_0 in the form

$$I_0 = \{a + ib \in \mathbb{Z}[i] \ : \ a \equiv b \bmod 2\} = (1 + i)\mathbb{Z}[i] = (\alpha),$$

it occurs that I_0 is the principal ideal in the ring of integers $\mathbb{Z}[i]$ generated by α. In fact, this ideal is identical to the set of possible partial quotients in Julius Hurwitz's case (Fig. 31).

To construct a corresponding continued fraction expansion, firstly an appropriate complex analogue to the Gaussian brackets is required. To identify the nearest integer to z from I_0, we define

$$[\,.\,]_T : \mathbb{C} \to I_0,$$

$$[z]_T = \left\lfloor x + \frac{1}{2} \right\rfloor \alpha + \left\lfloor y + \frac{1}{2} \right\rfloor \overline{\alpha}.$$

Furthermore, we specify a fundamental domain

$$X = \left\{ z = x\alpha + y\overline{\alpha} \ : \ \frac{-1}{2} \le x, y < \frac{1}{2} \right\},$$

and define a map $T : X \to X$ by $T0 = 0$ and

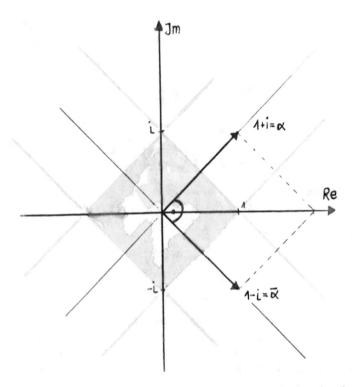

Fig. 31 Illustration, taken from [36], of Tanaka's change of coordinates $\{1, i\} \rightarrow \{\alpha, \overline{\alpha}\}$ along the angle bisectors of the first and third quadrant

$$Tz = \frac{1}{z} - \left[\frac{1}{z}\right]_T \quad \text{for } z \neq 0.$$

By iteration, we obtain the expansion

$$z = \frac{1}{a_1 + Tz} = \cfrac{1}{a_1 + \cfrac{1}{a_2 + T^2 z}} = \cdots = \cfrac{1}{a_1 + \cfrac{1}{a_2 + \ddots + \cfrac{1}{a_n + T^n z}}},$$

with partial quotients $a_n := a_n(z) := \left[\frac{1}{T^{n-1}z}\right]_T$. As in the real case, convergents $\frac{p_n}{q_n}$ to z are defined by "cutting" the continued fraction after the nth partial quotient. Setting

$$p_{-1} = \alpha, \quad p_0 = 0, \quad \text{and} \quad p_{n+1} = a_{n+1}p_n + p_{n-1} \quad \text{for } n \geq 1, \tag{4.3}$$

as well as

$$q_{-1} = 0, \quad q_0 = \alpha, \quad \text{and } q_{n+1} = a_{n+1}q_n + q_{n-1} \text{ for } n \geq 1, \tag{4.4}$$

by the same reasoning as for regular continued fractions for real numbers, z can be expressed as

$$z = \frac{p_n + T^n z p_{n-1}}{q_n + T^n z q_{n-1}}. \tag{4.5}$$

4.1.1 Example

As illustrating example we choose $x = \frac{1}{8}, y = \frac{3}{8}$ and receive

$$z = \frac{1}{2}(1 + i) - \frac{1}{4}(1 - i) = \frac{1}{8} + i\frac{3}{8} = \frac{1}{8}(1 - 3i).$$

For the first transformation Tz, we calculate $\frac{1}{z}$ and $\left[\frac{1}{z}\right]_T$:

$$\frac{1}{z} = \frac{8}{1 - 3i} = \frac{8(1 + 3i)}{(1 - 3i)(1 + 3i)} = \frac{8}{10}(1 + 3i) = \frac{4}{5}(1 + 3i).$$

Following Tanaka, we obtain

$$\binom{x}{y} = \frac{1}{2}\begin{pmatrix} 1 & 1 \\ 1 & -1 \end{pmatrix} \cdot \begin{pmatrix} \frac{4}{5} \\ \frac{12}{5} \end{pmatrix}$$

respectively $x = \frac{8}{5}, y = -\frac{4}{5}$. Thus, we have

$$a_1 = \left[\frac{1}{z}\right]_T = \left\lfloor \frac{8}{5} + \frac{1}{2} \right\rfloor \alpha + \left\lfloor -\frac{4}{5} + \frac{1}{2} \right\rfloor \overline{\alpha} = 2\alpha - \overline{\alpha} = 1 + 3i.$$

Consequently, we receive the continued fraction expansion with first partial quotient

$$z = \frac{1}{8}(1 - 3i) = \frac{1}{\frac{1}{z}}$$

$$= \frac{1}{\frac{4}{5}(2\alpha - \overline{\alpha})} = \frac{1}{2\alpha - \overline{\alpha} + \frac{4}{5}(2\alpha - \overline{\alpha}) - (2\alpha - \overline{\alpha})} = \frac{1}{a_1 + Tz}.$$

For the second partial quotient, we calculate

$$Tz = \frac{4}{5}(2\alpha - \overline{\alpha}) - (2\alpha - \overline{\alpha}) = -\frac{2}{5}\alpha + \frac{1}{5}\overline{\alpha} = -\frac{1}{5} - \frac{3}{5}i = -\frac{1}{5}(1 + 3i),$$

$$\frac{1}{Tz} = \frac{-5(1 - 3i)}{(1 + 3i)(1 - 3i)} = \frac{-5(1 - 3i)}{10} = -\frac{1}{2}(1 - 3i) = \frac{1}{2}(\alpha - 2\overline{\alpha})$$

and

$$a_2 = \left[\frac{1}{Tz}\right]_T = \left[\frac{1}{2} + \frac{1}{2}\right]\alpha + \left[-1 + \frac{1}{2}\right]\overline{\alpha} = \alpha - \overline{\alpha} = 2i.$$

For the third partial quotient, we calculate

$$T^2 z = -\frac{1}{2}(1 + i),$$

$$\frac{1}{T^2 z} = \frac{-2(1 - i)}{(1 + i)(1 - i)} = -1 + i = -\overline{\alpha}$$

and

$$a_3 = \left[\frac{1}{T^2 z}\right]_T = \left[-1 + \frac{1}{2}\right]\overline{\alpha} = -1 + i.$$

Since

$$T^3 z = \frac{1}{T^2 z} - \left[\frac{1}{T^2 z}\right]_T = 0,$$

it follows that the algorithm terminates and that the finite continued fraction is given by

$$\frac{1 - 3i}{8} = \cfrac{1}{1 + 3i + \cfrac{1}{2i + \cfrac{1}{-1 + i}}}.$$

4.1.2 Some Considerations and Characteristics

To deepen the understanding of the behaviour of T we examine some of its characteristics generating the complex continued fraction expansion. First of all, we verify that T indeed maps X to X and determine the region in which $\frac{1}{z}$ is located.

Lemma 4.1 *For $z \in X$ we have $Tz \in X$.*

Proof Considering $0 \neq z = a + ib = \frac{1}{2}(a + b)\alpha + \frac{1}{2}(a - b)\overline{\alpha} \in X$, it follows that

$$\frac{1}{z} = \frac{(a - b)}{2(a^2 + b^2)}\alpha + \frac{(a + b)}{2(a^2 + b^2)}\overline{\alpha}$$

and

$$\left[\frac{1}{z}\right]_T = \left[\frac{a^2 + a - b + b^2}{2(a^2 + b^2)}\right]\alpha + \left[\frac{a^2 + a + b + b^2}{2(a^2 + b^2)}\right]\overline{\alpha}.$$

To show that $Tz = \frac{1}{z} - \left[\frac{1}{z}\right]_T$ lies in X one takes the difference of the respective x or y-values. We receive

$$-\frac{1}{2} \le \frac{(a \pm b)}{2(a^2 + b^2)} - \left[\frac{a^2 + a \pm b + b^2}{2(a^2 + b^2)}\right] \le \frac{1}{2},$$

which concludes the proof. q.e.d.

In view of Lemma 4.1 the continued fraction expansion follows from iterating T over and over again. Since $Tz \in X$, we can localize $\frac{1}{z}$ by inverting the minimal and maximal possible value of $|z|$.

A natural question is:

For which numbers does the algorithm terminate?

To tackle this question we introduce a certain dissection of the set of Gaussian integers on behalf of

$$J := 1 + (\alpha) = \{c + id \in \mathbb{Z}[i] \ : \ c \not\equiv d \bmod 2\},$$

which leads to the disjoint decomposition

$$\mathbb{Z}[i] = I_0 \mathbin{\dot{\cup}} J$$

with I_0 as defined in Sect. 4.1. Recall that $I = I_0 \setminus \{0\}$ (Fig. 32).

Theorem 4.2 *The algorithm terminates if, and only if,*

$$z \in Q := \left\{z = \frac{u}{v} \ : \ \text{either } u \in I, v \in J \text{ or } v \in I, u \in J\right\}.$$

In the sequel the set

$$M := \{z \in \mathbb{C} \ : \ T^n z = 0 \text{ for some } n \in \mathbb{N}\} \subset \mathbb{Q}[i]$$

will be examined. Since the algorithm terminates if, and only if, z is equal to a convergent $\frac{p_n}{q_n}$, M can also be written as

$$M := \left\{z = \frac{p_n}{q_n} \text{ for some } n \in \mathbb{N}\right\}.$$

The proof of the theorem will be separated into two parts. After showing $M \subset Q$, we verify the inverse inclusion $Q \subset M$. However, first some preliminary work needs to be done.

By definition $I_0 = \{n\alpha + m\bar{\alpha} \ : \ m, n \in \mathbb{Z}\} = (1 + i)\mathbb{Z}[i] = (\alpha)$. Hence, for $z \in M$, we have

$$a_n = \left[\frac{1}{T^{n-1}z}\right]_T = \alpha b_n \text{ with } b_n \in \mathbb{Z}[i].$$

In accordance with (4.3) and (4.4) the first convergents are of the form

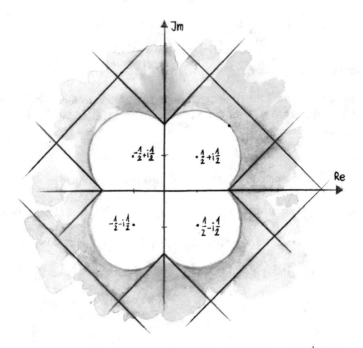

Fig. 32 Illustration taken from [36]: considerations from Sect. 2 show that $\frac{1}{z}$ is always located in the infinite region outside the four semicircles $B_{\pm 1 \pm i}$ centered at $\pm \frac{1}{2} \pm i \frac{1}{2}$ with radius $\frac{1}{\sqrt{2}}$ (see Sect. 4.1.3 for more details)

$$\frac{p_1}{q_1} = \frac{\alpha}{a_1 \alpha} = \frac{\alpha}{\alpha^2 b_1} \ , \quad \frac{p_2}{q_2} = \frac{a_2 \alpha}{(a_2 a_1 + 1)\alpha} = \frac{\alpha^2 b_2}{\alpha b'_2} \ , \quad \cdots$$

with $b'_2 \in \mathbb{Z}[i]$ and so on. We observe that the parity of the α-parts of numerator and denominator alternate in their exponents.

Lemma 4.3 *We have*

$$\frac{p_n}{\alpha} \in (\alpha) \quad \Leftrightarrow \quad 2 | n \quad \Leftrightarrow \quad \frac{q_n}{\alpha} \notin (\alpha),$$

respectively

$$\frac{p_n}{\alpha} \notin (\alpha) \quad \Leftrightarrow \quad 2 \nmid n \quad \Leftrightarrow \quad \frac{q_n}{\alpha} \in (\alpha).$$

Proof It is sufficient to consider the sequence of nominators (p_n), since (q_n) can be treated analogously. According to the recursion formula (4.3) we may write

$$\frac{p_n}{\alpha} = a_n \frac{p_{n-1}}{\alpha} + \frac{p_{n-2}}{\alpha},$$

what leads to a simple proof by induction. We begin with

- $n = -1 : \frac{p_{-1}}{\alpha} = \frac{\alpha}{\alpha} = 1 \notin (\alpha)$, and
- $n = 0 : \frac{p_0}{\alpha} = 0 \in (\alpha)$,

which satisfy the assertion. In the induction step we distinguish the two cases of n being even or odd.

1. *case* : $2 | n$
   Since $n - 2$ is even as well, the induction hypothesis provides $\frac{p_{n-2}}{\alpha} \in (\alpha)$. Moreover, we observe $a_n \in (\alpha)$, respectively $a_n \frac{p_{n-1}}{\alpha} \in (\alpha)$. On behalf of the recursion formula follows $\frac{p_n}{\alpha} \in (\alpha)$. Here we have used that (α) is an ideal.
2. *case* : $2 \nmid n$
   This case can be treated similarly: $\frac{p_{n-2}}{\alpha} \notin (\alpha)$, $a_n \in (\alpha)$ implies $\frac{p_n}{\alpha} \notin (\alpha)$.

q.e.d.

Thus, we have verified $M \subset Q$ and for the proof of Theorem 4.2 it remains to prove the inverse inclusion.

Lemma 4.4 *We have*

$$Q \subset M.$$

In particular, all $z = \frac{u}{v} \in \mathbb{Q}(i)$ with either $u \in I, v \in J$ or $v \in I, u \in J$ lead to a terminating algorithm.

Proof We assume that $O := Q \setminus M$ is not an empty set. Then there exists a number $z \in O$ of the form $z = \frac{u}{v} \in \frac{I}{J}$ (respectively $\in \frac{J}{I}$) with $T^n \frac{u}{v} \neq 0$, for which $|u|^2 + |v|^2$ is minimal. Without loss of generality we may assume $u \in I$ and $v \in J$.
We have

$$z = \frac{u}{v} = \cfrac{1}{a_1 + \cfrac{1}{a_2 + \cfrac{1}{\ddots + \cfrac{1}{a_n + T^n \frac{u}{v}}}}}.$$

Furthermore, we put

$$z' = \frac{x}{y} = \cfrac{1}{a_2 + \cfrac{1}{a_3 + \cfrac{1}{\ddots + \cfrac{1}{a_n + T^n \frac{u}{v}}}}}.$$

Because of $T^n \frac{u}{v} = T^{n-1} \frac{x}{y}$, it follows that z' is as well an element of the set Q and satisfies the condition

$$T^n \frac{x}{y} \neq 0.$$

Moreover, we have

$$\frac{u}{v} = \frac{1}{a_1 + \frac{x}{y}} = \frac{y}{a_1 y + x} \tag{4.6}$$

and therewith $u = y \in I$ (respectively $\in J$) and $v = a_1 y + x \in J$ (respectively $\in I$). To go on, we observe the multiplicative and additive structure of I and J by noting

$$J \cdot I = I \cdot J = I, J \cdot J = J, I \cdot I = I,$$

and

$$J + I = I + J = J, J + J = I, J + J = J.$$

Of course, this is meant in the sense of Minkowski's multiplication and addition of sets, respectively, and can easily be shown on behalf of complex calculation.

With the constraint $a_1 \in I$, the structure of $\frac{x}{y}$ occurs:

1. For $u = y \in I$ and $v = a_1 y + x \in J$ we have that $a_1 y + x$ is an element of $I \cdot I + x$, which is a subset of J, if, and only if, $x \in J$.
2. For $u = y \in J$ and $v = a_1 y + x \in I$ we have that $a_1 y + x$ is an element of $I \cdot J + x$, which is a subset of I, if, and only if, $x \in I$.

Obviously, the structure of $z = \frac{u}{v}$ changes for $z' = \frac{x}{y}$ in an alternating way. The property of non-terminating as well as the observed structure above do also hold if $z = \frac{x}{y}$ is replaced by $\frac{x}{-y}$. With (4.6) we additionally obtain

$$|u| = |y| \text{ and } |v| = |a_1 y + x|$$

what leads to

$$|u|^2 + |v|^2 = |y|^2 + |a_1 y + x|^2.$$

Since $|u|^2 + |v|^2$ was assumed to be minimal, the inequality

$$|a_1 y + x|^2 \le |x|^2 \tag{4.7}$$

follows. Since setting $z' = \frac{x}{y}$ or $z' = \frac{x}{-y}$ will not change anything of our previous observations, it is guaranteed that the case of real part and imaginary part of $a_1 y$ being positive can always be achieved. With $x = a + ib$, the inequality

$$|a_1 y + x|^2 = (\text{Re}(a_1 y) + a)^2 + (\text{Im}(a_1 y) + b)^2 > a^2 + b^2 = |x|^2$$

holds. This is a contradiction to (4.7), i.e., to the minimality of $z = \frac{u}{v}$, which proves that there is no number satisfying the assumption. Consequently, O is an empty set. q.e.d

Concluding Lemmas 4.3 and 4.4, it follows that complex numbers $z \in \mathbb{C}$, having a finite continued fraction, are either of the form $z \in \frac{I}{J}$ or $z \in \frac{J}{I}$.

Remark In Julius Hurwitz's doctoral thesis [30] a whole chapter is dedicated to the question which complex numbers have a finite continued fraction expansion. There it is described that the continued fraction is finite in any case of rational complex numbers. This is because Hurwitz allowed a last partial quotient from $\mathbb{Z}[i]$ in those cases when irregularities occur from rational complex numbers with numerator and denominator divisible by $(1 + i)$. That Shigeru Tanaka [46] did not refer to those difficulties is certainly due to the fact that under an ergodic theoretical point of view those few irregularities are insignificant. The set $\mathbb{Q}(i)$ is countable and thus negligible with respect to applications from ergodic theory. However, this could also indicate that Julius Hurwitz's thesis was unknown to Tanaka.

Next we examine some characteristics of the algorithm concerning its approximation property.

Lemma 4.5 *Let $a_{n+1} \neq 0$. For $k_n := \frac{q_{n+1}}{q_n}$ we have $|k_n| > 1$.*

Notice that for $a_{n+1} = 0$ the recursion formula leads to $q_{n+1} = q_{n-1}$ and that the continued fraction is finite.

Proof We suppose that all previous $k_1, \ldots, k_{n-1}$ are of absolute value > 1, here certainly $|k_1| \geq \sqrt{2}$ (because $|a_j| \geq \sqrt{2}$ for every $a_j \in I$). Thus,

$$k_n = a_{n+1} + \frac{1}{k_{n-1}} \in \{z \in \mathbb{C} : |z - a_n| < 1\}.$$

We assume $|k_n| < 1$, hence,

$$|a_{n+1}| = \left| k_n - \frac{1}{k_{n-1}} \right| \leq |k_n| + \frac{1}{|k_{n-1}|} < 2$$

and consequently $a_{n+1} = \pm 1 \pm i$. Without loss of generality we consider $a_{n+1} = 1 + i$. By a backwards calculation we determine

$$a_{n-2k} = -2 + 2i \quad \text{and} \quad a_{n-2k-1} = 2 + 2i$$

for all $k \in \mathbb{N}_0$ with $2k < n$. The following considerations are by induction:
For $n = 1$ we have $k_1 = a_2 = 1 + i$ which leads to $|k_1| = \sqrt{2} > 1$.
For $n = 2$ we have $k_2 = a_3 + \frac{1}{-2+2i} = \frac{3}{4}(1 + i)$ which leads to $|k_1| = \frac{3}{4}\sqrt{2} > 1$.
For $n \geq 3$ we have $k_1 = \pm 2 + 2i$ which leads to $|k_1| = \sqrt{8}$ and $k_2 = \pm 2 + 2i + \frac{1}{k_1}$ which leads to $|k_2| \geq \sqrt{8} - \frac{1}{|k_1|}$.
Therefore, we observe $k_j = \pm 2 + 2i + \frac{1}{k_{j-i}}$ as well as $|k_j| \geq \sqrt{8} - \frac{1}{|k_{j-1}|} =: y_j \in \mathbb{R}$.
We consider the recursion $y_j = \sqrt{8} - \frac{1}{y_{j-1}}$ with

$$y_1 = \sqrt{8}, \; y_2 = \sqrt{8} - \frac{1}{\sqrt{8}} = \frac{7}{4}\sqrt{2} < \sqrt{2} + 1.$$

Under the assumption $y_j < y_{j-1} < \ldots$ it follows by induction that

$$y_{j+1} = \sqrt{8} - \frac{1}{y_j} < \sqrt{8} - \frac{1}{y_{j-1}} = y_j$$

as well as

$$y_{j+1} = \sqrt{8} - \frac{1}{y_j} > \sqrt{8} - \frac{1}{\sqrt{2}+1} = \sqrt{2}+1.$$

Hence, for all $j < n$ the inequality

$$k_j \geq y_j > \sqrt{2}+1$$

holds. Consequently, with $k_n = a_n + \frac{1}{k_{n-1}}$ this leads to

$$|k_n| = \left| 1 + i + \frac{1}{k_{n-1}} \right| > \sqrt{2} - \frac{1}{\sqrt{2}+1} = 1.$$

This is the desired inequality. q.e.d.

In the following, we shall use the previous lemma in order to show that the continued fraction expansion actually converges.

4.1.3   Geometrical Approach to the Approximation Behaviour

We have already indicated that, following Tanaka's line of argumentation, the approximation properties of his algorithm can be illustrated geometrically. This happens essentially on behalf of two tilings.

Firstly, the fundamental domain X is split into disjoint so-called T-cells. This is done with respect to the first n partial quotients as follows. We define the set $A(n)$ of sequences of partial quotients by

$$A(n) = \{a_1(z), a_2(z), \ldots, a_n(z) : z \in X\}.$$

Such sequences are called T-admissible. One should notice that certain sequences of numbers from $(1+i)\mathbb{Z}[i]$ cannot appear as sequence of partial quotients, which was examined by Julius Hurwitz (see Lemma 2.1).

Corresponding to each admissible sequence $a_1, a_2, \ldots, a_n \in A(n)$ the subset $X(a_1, a_2 \ldots, a_n)$ of X arises as

$$X(a_1, a_2 \ldots, a_n) = \{z \in X : a_k(z) = a_k \text{ for } 1 \leq k \leq n\};$$

each such set is called a T-cell. We have

$$X = \bigcup_{a_1, a_2 \ldots, a_n \in A(n)} X(a_1, a_2 \ldots, a_n).$$

In other words, T-cells describe a "close" neighborhood of a certain $z \in X$.

The second tiling is corresponding to the set of reciprocals

$$X^{-1} = \left\{ \frac{1}{z} : z \in X, z \neq 0 \right\}.$$

In each iteration of the continued fraction algorithm a complex number

$$T^n z = \frac{1}{T^{n-1}z} - \left[\frac{1}{T^{n-1}t} \right]_T$$

arises, which is split into an integral partial quotient and a remainder. The latter is going to be iterated again. We thus receive a sequence of elements of X^{-1}. By taking the reciprocals, the edges of X are transformed to arcs of discs. We define

$$U_1 := \left\{ z \in X : \left| z + \frac{\alpha}{2} \right| \geq \frac{1}{\sqrt{2}} \right\}.$$

Writing $z = x\alpha + y\overline{\alpha}$, we have

$$\left| z + \frac{\alpha}{2} \right| = \left| \left(x + y + \frac{1}{2} \right) + i \left(x - y + \frac{1}{2} \right) \right|$$

$$= \sqrt{ \left(\mathrm{Re}\,(z) + \frac{1}{2} \right)^2 + \left(\mathrm{Im}\,(z) + \frac{1}{2} \right)^2 } \geq \frac{1}{\sqrt{2}}$$

including elements $z \in X$, while excluding numbers which lie inside the disc of radius $\frac{1}{\sqrt{2}}$ and center $-\frac{1}{2}(1 + i) = -\frac{1}{2}\alpha$ (see Fig. 32). Analogously, we define

$$U_2 := -i \times U_1, \quad U_3 := -i \times U_2 \text{ and } U_4 := -i \times U_3.$$

Setting
$$U(\alpha) := U_1, \quad U(\overline{\alpha}) := U_2, \quad U(-\alpha) := U_3, \quad U(-\overline{\alpha}) := U_4$$

and
$$U(a) := X, \text{ if } a \neq \alpha, \overline{\alpha}, -\alpha, -\overline{\alpha},$$

we attach to each Gaussian integer $a \in I$ an area. With (4.2) there is always an even integer distance between real part $\mathrm{Re}\,a$ and imaginary part $\mathrm{Im}\,a$ (Fig. 33).

Thus, the set of reciprocals X^{-1} can be composed from translates of the sets $U(a)$ shifted by a, i.e.,

$$X^{-1} = \bigcup_{a \in I} (a + U(a)).$$

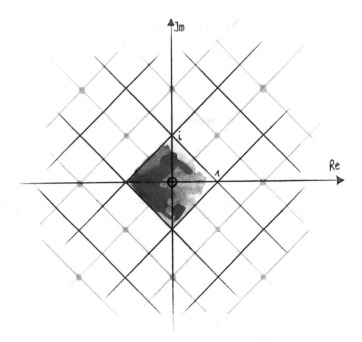

Fig. 33 Illustration taken from [36]: the numbers $a \in I$ are located on the angle bisectors of the quadrants or on parallel lines shifted by a multiple of 2 excluding the origin

These geometrical observations are related to one another on behalf of the defined transformation T. We have

$$T^n X(a_1, \ldots, a_n) = U(a_n).$$

This implies that the nth iteration of the transformation applied to the remainder of the n first partial quotients maps to the domain U, which is located around the nth partial quotient. This is interesting in view of Eq. (4.5), from which the uniqueness of the inverse map T^n follows. We define $\phi_{a_1 \cdots a_n} := (T^n)^{-1}$ by

$$\phi_{a_1 \cdots a_n} : U(a_n) \to X(a_1, \ldots, a_n)$$

with

$$\phi_{a_1 \cdots a_n}(z) = \frac{p_n + z p_{n-1}}{q_n + z q_{n-1}}.$$

The "forward" mapping to the non-integer remainder of $\frac{1}{T^{n-1}z}$ becomes a unique inverse mapping "backwards" in the algorithm. Since in each set $U(a_n) = X, U_1, \ldots,$ U_4 the origin is included, for all admissible sequences of partial quotients $a_1, \ldots, a_n \in A(n)$ the nth convergent is located in the corresponding T-cell, that is

$$\frac{p_n}{q_n} = \phi_{a_1 \cdots a_n}(0) \in X(a_1 \cdots a_n).$$

Consequently, the algorithm indeed produces convergents approximating the initial values better and better. The more partial quotients are chosen as fixed, the smaller the T-cell becomes in X and therewith the closer the corresponding convergent is located.

4.2 Ergodic Theory

In ergodic theory one examines so-called *measure preserving dynamical systems*. In general such a dynamical system describes a mathematical concept which models a certain time-process in a certain space on behalf of fixed mathematical regularities.

4.2.1 Transformations

We consider a probability space (X, Σ, μ) with non-empty set X, a σ-algebra Σ on the set X, a probability measure μ on (X, Σ) and a measure preserving *transformation*

$$T : X \to X.$$

The above mentioned time-process is explained by assuming T as "shift into the future", whereas its inverse T^{-1} can be considered as "shift into the past". Here T is said to be *measure preserving*, if, and only if, for $E \in \Sigma$,

$$\mu(T^{-1}E) = \mu(E),$$

which means that the measure of E is preserved under T. Furthermore, a measurable set E is called T-*invariant* when

$$T^{-1}E = E.$$

The corresponding dynamical system is written as quadrupel (X, Σ, μ, T).

In addition, T is called *ergodic* when for each μ-measurable, T-invariant set E either

$$\mu(E) = 0 \text{ or } \mu(E) = 1.$$

A very powerful result dealing with ergodic transformations was realized by Georg David Birkhoff [2] and in the general form below by Aleksandr Jakovlevich Khinchine [33].[74]

---

[74]For a comprehensive version see [7].

Theorem 4.6 (Pointwise Ergodic Theorem, Birkhoff 1931)
Let T be a measure preserving ergodic transformation on a probability space (X, Σ, μ). If f is integrable, then

$$\lim_{N \to \infty} \frac{1}{N} \sum_{0 \leq n \leq N} f(T^n x) = \int_X f d\mu$$

for almost all $x \in X$.

Here the nth transformation T^n is defined recursively as follows

$$T^0 = id, \quad T^1 = T \text{ and } T^n = T^0 \circ T^{n-1}.$$

4.2.2 Continued Fraction Transformation

In this subsection we shortly introduce an ergodic approach to continued fractions. Hereby, we follow a classical method[75] firstly explained for the real case.

We define a transformation $T : [0, 1) \to [0, 1)$ which serves as operator in the well known regular continued fraction algorithm (similar to the map T introduced in Tanaka's complex approach in Sect. 4.1). Choosing the unit interval as fundamental set X, we define $T0 = 0$ and, for $x \in X$,

$$Tx = T(x) := \frac{1}{x} - \left\lfloor \frac{1}{x} \right\rfloor \text{ if } x \neq 0.$$

For an irrational number $x \in \mathbb{R} \setminus \mathbb{Q}$ and $a_0 := \lfloor x \rfloor \in \mathbb{Z}$ we obviously have $x - a_0 \in [0, 1)$. Setting

$$T^0 x := x - a_0, \quad T^1 x := T(x - a_0), \quad T^2 x := T(T^1 x), \ldots,$$

the definition above provides

$$T^n x \in [0, 1) \setminus \mathbb{Q} \text{ for all } n \geq 0.$$

With

$$a_n = a_n(x) := \left\lfloor \frac{1}{T^{n-1} x} \right\rfloor \text{ for } n \geq 1$$

one receives the well-known regular continued fraction expansion

---

[75] Once more we refer to [7, p. 20].

$$x = a_0 + \cfrac{1}{a_1 + T^1 x} = a_0 \cfrac{1}{a_1 + \cfrac{1}{a_2 + T^2 x}} = \cdots$$

$$= a_0 + \cfrac{1}{a_1 + \cfrac{1}{a_2 + \cdots + \cfrac{1}{a_n + T^n x}}} = [a_0; a_1, a_2, \cdots, a_n + T^n x].$$

The existence of the limit

$$x = [a_0; a_1, a_2 \cdots, a_n, \cdots] = \lim_{n \to \infty} [a_0; a_1, a_2 \cdots, a_n + T^n x]$$

follows on behalf of the representation related to the nth convergent $\frac{p_n}{q_n} \in \mathbb{Q}$ to x. In fact, we have

$$x = \frac{p_n + T^n x p_{n-1}}{q_n + T^n x q_{n-1}}$$

and $p_{n-1} q_n - p_n q_{n-1} = (-1)^n$ for $n \geq 1$. In view of $T^n x \in [0, 1)$ the inequality

$$\left| x - \frac{p_n}{q_n} \right| < \frac{1}{a_{n+1} q_n^2}$$

can be derived. Here $(q_n)_{n \geq 0}$ is a strictly increasing sequence of positive integers.

In the complex case, the fundamental set is naturally two-dimensional corresponding to real and imaginary parts of the expanded complex number. Hence, in Shigeru Tanaka's, respectively Julius Hurwitz's algorithm the transformation $T : X \to X$ is defined on the fundamental domain $X = \{z = x\alpha + y\overline{\alpha} : \frac{-1}{2} \leq x, y < \frac{1}{2}\}$ with $\alpha = 1 + i$ (and $\overline{\alpha} = 1 - i$) by

$$Tz := \frac{1}{z} - \left[\frac{1}{z} \right]_T \quad \text{for } z \neq 0 \quad \text{and} \quad T0 = 0,$$

where

$$[z]_T := \left\lfloor x + \frac{1}{2} \right\rfloor (1 + i) + \left\lfloor y + \frac{1}{2} \right\rfloor (1 - i).$$

The analogy between the real and complex approach is obvious. However, for the complex continued fractions some adjustments need to be done. In order to apply ergodic methods a main challenge is to define an invariant measure for a given transformation. In the real case, there is the so-called *Gauss-measure*,[76] defined by

$$\mu(A) = \frac{1}{\log 2} \int_A \frac{1}{1+x} dx,$$

---

[76] Discovered by Carl Friedrich Gauss.

for all Lebesgue sets $A \subset [0, 1)$. For complex algorithms some difficulties may appear. Here a so-called natural extension of the underlying transformation can be helpful.

4.2.3 Dual Transformation and Natural Extension for Tanaka's Algorithm

We have shown that Tanaka's transformation T provides a complex continued fraction expansion. Here we sketch another related transformation $S : Y \to Y$, introduced by Tanaka, where

$$Y = \{w \in \mathbb{C} : |w| \leq 1\}$$

is the unit disc centered at the origin in the complex plane. We define subsets V_j of Y by

$$V_1 := \{w \in Y : |w + \alpha| \geq 1\},$$

$$V_2 = -i \times V_1, \quad V_3 = -i \times V_2, \quad V_4 = -i \times V_3, \quad V_5 = V_1 \cap V_2,$$

$$V_6 = -i \times V_5, \quad V_7 = -i \times V_6, \quad V_8 = -i \times V_7$$

and a partition of $I_0 = \{n\alpha + m\bar{\alpha} \; : \; n, m \in \mathbb{Z}\}$, respectively $I = I_0 \setminus \{0\}$ by

$$J_1 = \{n\alpha : n > 0\}, J_2 = -i \times J_1, \quad J_3 = -i \times J_2, \quad J_4 = -i \times J_3,$$

$$J_5 = \{n\alpha + m\bar{\alpha} : m > 0\}, J_6 = -i \times J_5, \quad J_7 = -i \times J_6, \quad J_8 = -i \times J_7.$$

Setting

$$V(a) := \begin{cases} Y, & \text{if } a = 0, \\ V_j, & \text{if } a \in J_j, \end{cases}$$

for $1 \leq j \leq 8$, we obtain a new complete tiling of the complex plane

$$\mathbb{C} = \bigcup_{a \in I_0} (a + V(a)).$$

Furthermore, we define $S0 = 0$ and

$$Sw = \frac{1}{w} - \left[\frac{1}{w}\right]_S \quad \text{for } w \neq 0$$

with $[w]_S = a$ if $w \in a + V(a)$. As above partial quotients arise through

$$b_n = b_n(w) = \left[\frac{1}{S^{n-1}w}\right]_S \in I_0.$$

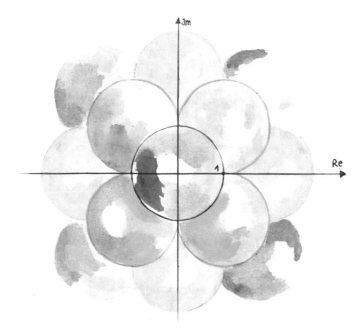

Fig. 34 Illustration of the tiling of the complex plane with respect to S

This leads to an expansion of $w \in Y$ as

$$w = [b_1, b_2, \cdots, b_n + S^n w]$$

with convergents $V_n := [b_1, b_2, \cdots, b_n]$. Tanaka proved that the transformation S satisfies a certain duality.

Lemma 4.7 (Duality)
Let $a_1, \ldots, a_n$ be a sequence of numbers in I. Then $a_1, \ldots, a_n$ is T-admissible if, and only if, the reverse sequence of partial quotients $a_n, \ldots, a_1$ is S-admissible.

The proof is based on geometrical constraints concerning a finely tiling of Y respectively Y^{-1} and can be found in [46, p. 200] (Fig. 34).

By Lemma 4.7, we know that for each T-admissible sequence there exists an associated sequence of convergents of a given $w \in Y$, namely

$$\frac{q_{n-1}}{q_n} = [a_n, a_{n-1}, \ldots, a_1] = V_n \in Y.$$

This provides another proof of Lemma 4.5: the sequence $(q_n)_{n \in \mathbb{N}}$ increases monotonously in absolute value. Given this *dual transformation* S, one can construct the so-called *natural extension* T containing information of T as well as information of S, which can be regarded as "future" and "past" of the sequence of partial quotients.

On behalf of Lemma 4.7, we define $\mathsf{T} : X \times Y \to X \times Y$ by[77]

$$\mathsf{T}(z, w) = \left(\mathsf{T}z, \frac{1}{a_1(z) + w} \right), \tag{4.8}$$

where $a_1 = \left[\frac{1}{z} \right]_T$. Tanaka [46] proved.

Theorem 4.8 (Natural Extension)
The transformation T is a natural extension; in particular T and S are ergodic and the function $h : X \times Y \to \mathbb{R}$ defined by

$$h(z, w) = \frac{1}{|1 + zw|^4}$$

is the density function of a finite absolutely continuous T-invariant measure.

4.2.4 Variation on the Doeblin-Lenstra Conjecture

In this section we give a proof of an analogue of the so-called Doeblin-Lenstra Conjecture for the complex case of Tanaka's continued fraction algorithm [37]. Therefore, we first state the original conjecture for the regular continued fraction algorithm for real numbers.

Theorem 4.9 (Doeblin-Lenstra Conjecture)
Let x be any irrational number with continued fraction convergents $\frac{p_n}{q_n}$. Define the approximation coefficients by $\theta_n := q_n|q_n x - p_n|$. Then for almost all x we have

$$\lim_{n \to \infty} \frac{1}{n} \left| \{ j \ : \ j \le n, \theta_j(x) \le c \} \right| = \begin{cases} \frac{c}{\log 2}, & \text{for } 0 \le c \le \frac{1}{2}, \\ \frac{-c + \log 2c + 1}{\log 2}, & \text{for } \frac{1}{2} \le c \le 1. \end{cases} \tag{4.9}$$

Notice that $\theta_j(x) \le 1$ for all j and all x.

Indeed, this conjecture is due to Hendrik W. Lenstra jr. and, implicitly, included in an old paper of Wolfgang Doeblin [8]; it has been proven by Wieb Bosma, Hendrik Jager and Freek Wiedijk [4] with ergodic methods, in particular by using the natural extension T of the continued fraction transformation T as main tool. We shall follow their approach and apply their machinery to Tanaka's, respectively Julius Hurwitz's complex continued fraction. For this purpose we define, for some complex number z,

$$\theta_n := \theta_n(z) := |q_n|^2 \left| z - \frac{p_n}{q_n} \right|.$$

---

[77]In [46] a more exact, more finely, tiling of X resp. Y was given on behalf of sets U and V. For our needs, it is sufficient to consider the whole sets X and Y.

Notice that the absolute values of the denominators q_n increase strictly with n. We shall prove the almost sure existence of a limiting distribution function for this quantity:

Theorem 4.10 *For $x > 0$ we define*

$$\ell(x) := \lim_{N \to \infty} \frac{1}{N} \sharp \{ n \le N : \theta_n(z) \le x \}.$$

Then, for almost all z, the distribution function $\ell(x)$ exists and is given by

$$\ell(x) = \lim_{N \to \infty} \frac{1}{N} \sum_{n \le N} \chi_{A_x}(\mathsf{T}^n(z, 0)) = \mu(A_x) = \frac{1}{G} \iint_{A_x} \frac{d\lambda(u, v)}{|1 + uv|^4}, \tag{4.10}$$

where T is defined by (4.12) below, λ is the Lebesgue measure, χ_{A_x} is the characteristic function of the set

$$A_x := \left\{ (u, v) \in X \times Y : \left| \frac{1}{u} + v \right| \ge \frac{1}{x} \right\},$$

and the normalizing constant

$$G := \iint_{X \times Y} \frac{d\lambda(u, v)}{|1 + uv|^4} = 4\pi \sum_{n=0}^{\infty} \frac{(-1)^n}{(2n + 1)^2}. \tag{4.11}$$

equals 4π times the Catalan constant.

Interestingly, Catalan's constant appears in various results in the ergodic theory of continued fractions, see for example [34], [32, p. 1217] or [9, p. 2].

We shall briefly mention another interesting phenomenon: already in Julius Hurwitz's thesis [30] one can find the statement that the sequence of $\theta_n(z)$ has always a finite limit point (p. 34), in general, however, the sequence of the $\theta_n(z)$ appears to have divergent subsequences. This is rather different from the real case where the limiting distribution function is constant for $x \ge 1$ as follows from the classical estimate $|z - \frac{p_n}{q_n}| \le \frac{1}{q_n^2}$ giving $\theta_n(z) \le 1$. Indeed, there exist z for which $\theta_n(z)$ diverges to infinity.

Proof of Theorem 4.10. Recall that Tanaka showed that the map $\mathsf{T} : X \times Y \to X \times Y$ defined by

$$\mathsf{T}(z, w) = \left(\mathsf{T}z, \frac{1}{a_1(z) + w} \right), \tag{4.12}$$

where $a_1 = \left[\frac{1}{z} \right]_T$ is a natural extension (Theorem 4.10); in particular T and S are ergodic. Moreover, the function $h : X \times Y \to \mathbb{R}$ given by

$$h(u, v) = \frac{1}{|1 + uv|^4} \tag{4.13}$$

is the density function of a finite absolutely continuous T-invariant measure μ. It thus follows from the dual transformation that

$$v_n = \frac{q_{n-1}}{q_n} = [0; a_n, a_{n-1}, \ldots, a_2, a_1],$$

and hence we obtain the equivalence

$$\theta_n(z) \leq x \quad \Leftrightarrow \quad \left| \frac{T^n z}{1 + V_n T^n z} \right| \leq x. \tag{4.14}$$

By (4.12), we find

$$T^n(z, w) = (T^n z, [0; a_n, a_{n-1}, \ldots, a_2, a_1 + w]),$$

and, in particular, $T^n(z, 0) = (T^n z, V_n)$. In view of (4.14) this shows that $\theta_n(z) \leq x$ if, and only if,

$$T^n(z, 0) \in A_x = \{(u, v) \in X \times Y : \left| \frac{u}{1 + uv} \right| \leq x\},$$

respectively

$$T^n(z, 0) \in A_x = \left\{ (u, v) \in X \times Y : \left| \frac{1}{u} + v \right| \geq \frac{1}{x} \right\}.$$

Comparing the quantities $T^n(z, w) = (T^n z, [0; a_n, a_{n-1}, \ldots, a_1 + w])$ and $T^n(z, 0) = (T^n z, [0; a_n, a_{n-1}, \ldots, a_1])$, it follows that for every $\epsilon > 0$ there exists $n_0(\epsilon)$ such that, for all $n \geq n_0(\epsilon)$ and all $w \in Y$, we have

$$T^n(z, w) \in A_{x+\epsilon} \Rightarrow T^n(z, 0) \in A_x,$$

as well as

$$T^n(z, 0) \in A_x \Rightarrow T^n(z, w) \in A_{x-\epsilon}.$$

We define $A_x^N := \{n \leq N : T^n(z, w) \in A_x\}$. Since T is ergodic, the following limits exist and the inequalities in between hold: for all $\epsilon > 0$,

$$\lim_{N \to \infty} \frac{1}{N} \sharp A_{x+\epsilon}^N \leq \liminf_{N \to \infty} \frac{1}{N} \sharp A_x^N \leq \limsup_{N \to \infty} \frac{1}{N} \sharp A_x^N \leq \lim_{N \to \infty} \frac{1}{N} \sharp A_{x-\epsilon}^N.$$

Since the underlying dynamical system $(X \times Y, \Sigma, \mu, T)$ with the corresponding invariant probability measure μ (with density given by (4.13)) and σ-algebra Σ is ergodic, application of Birkhoff's pointwise ergodic theorem yields (4.10).

In order to compute the normalizing factor G, we shall use the following derivation of the geometric series expansion

$$\frac{1}{(1-w)^2} = \sum_{m \geq 1} m w^{m-1},$$

valid for $|w| < 1$. To prevent difficulties that could arise from singularities of h on the boundary of $X \times Y$ we define, for $0 < \rho < 1$,

$$X_\rho := \{z = x + iy : -\frac{\rho}{2} \leq x, y \leq \frac{\rho}{2}\} \quad \text{and} \quad Y_\rho := \{w \in \mathbb{C} : |w| \leq \rho\}.$$

In view of (4.13) and $|1 + uv|^2 = (1 + uv)(1 + \overline{uv})$ we find

$$G_\rho := \iint_{X_\rho \times Y_\rho} \frac{d\lambda(u, v)}{|1 + uv|^4}$$

$$= \sum_{m,n \geq 1} mn(-1)^{m+n} \int_{X_\rho} u^{m-1} \overline{u}^{n-1} d\lambda(u) \int_{Y_\rho} v^{m-1} \overline{v}^{n-1} d\lambda(v).$$

We compute

$$\int_{Y_\rho} v^{m-1} \overline{v}^{n-1} d\lambda(v) = \int_{Y_\rho} |v|^{2(n-1)} v^{m-n} d\lambda(v)$$

$$= \int_0^{2\pi} \int_0^\rho r^{m+n-1} e^{i\varphi(m-n)} dr d\varphi = \begin{cases} \pi \frac{\rho^{2n}}{n}, & \text{if } m = n, \\ 0, & \text{otherwise.} \end{cases}$$

Applying the derivation of the geometric series expansion once again, leads to

$$G_\rho = \pi \sum_{n \geq 1} n \rho^{2n} \int_{X_\rho} |u|^{2(n-1)} d\lambda(u)$$

$$= \pi \int_{X_\rho} \sum_{n \geq 1} n \rho^{2n} |u|^{2(n-1)} d\lambda(u) = \pi \int_{X_\rho} \frac{d\lambda(u)}{\rho^{-2} - |u|^2}.$$

Now with $\rho \to 1-$ we get $G_\rho \to G_1 = G$ by Lebesgue's theorem on monotone convergence. Next, we use the symmetry of the fundamental domain and split it along the axes into four parts of equal size. We consider the part located in the first quadrant $\triangle := \{a + ib : 0 \leq a, b \leq 1; b - 1 < a\}$ and deduce

$$G = 4\pi \iint_\triangle \frac{da\,db}{1 - |a + ib|^2} = 4\pi \int_0^1 \left(\int_0^{1-a} \frac{db}{1 - (a^2 + b^2)} \right) da,$$

where

$$\int_0^{1-a} \frac{db}{1-a^2-b^2} = \int_0^{1-a} \left(\frac{1}{\sqrt{1-a^2}-b} + \frac{1}{\sqrt{1-a^2}+b} \right) db \frac{1}{2\sqrt{1-a^2}}$$

$$= \left[-\log|b - \sqrt{1-a^2}| + \log(b + \sqrt{1-a^2}) \right]_{b=0}^{1-a} \cdot \frac{1}{2\sqrt{1-a^2}}$$

$$= \frac{1}{2\sqrt{1-a^2}} \log \frac{\sqrt{1-a^2}+1-a}{\sqrt{1-a^2}-1+a}.$$

Altogether this gives

$$G = 2\pi \int_0^1 \log \frac{\sqrt{1-a^2}+1-a}{\sqrt{1-a^2}-1+a} \frac{da}{\sqrt{1-a^2}}$$

$$= 2\pi \left[A - B + i \sum_{k=1}^{\infty} \frac{(-\exp(i\arcsin(a)))^k}{k^2} - i \sum_{k=1}^{\infty} \frac{(\exp(i\arcsin(a)))^k}{k^2} \right]_{a=0}^{1}$$

with

$$A = \arcsin(a) \log(1 - \exp(i\arcsin(a))),$$
$$B = \arcsin(a) \log(1 + \exp(i\arcsin(a))).$$

Hence, we obtain (4.11).

In contrast to the real case for the ordinary regular continued fraction expansion [4] the analogous expressions for the complex situation are by far more complicated. Using Tanaka's results [46] one may compute $\ell(x)$ in the latter case numerically, however, it seems difficult to find an explicit expression for the limiting distribution function. Another instance of this difficulty is given by Tanaka [46] himself and his non-explicit respresentation of the entropy of T and S.

Appendix

The appendix contains two parts: a list of direct or indirect references to David Hilbert in Hurwitz's estate (see Sect. 3) in the ETH Zurich library and a table of figures of these notes.

Appendix I: Links to Hilbert in the ETH Estate of Hurwitz

In Sect. 3 we consider the teacher–student relation of Adolf Hurwitz and David Hilbert. Here we give a list of documents of Adolf Hurwitz's estate in the ETH Zurich library (in the directories HS 582 and HS 583) directly or indirectly connected to

David Hilbert including a great number of diary entries [21] (HS 582 : 1–30) with remarks related to Hilbert.

- Lectures of David Hilbert edited by Julius Hurwitz

  HS 582: 154, *Die eindeutigen Funktionen mit linearen Transformationen in sich* (Königsberg 1892 SS) with handwritten remarks of Hilbert (e.g. on pages −1, 34, 69, 80, 81, ...)

  HS 582: 158, *Geometrie der Lage* (Königsberg 1891 SS)

- No. 6: 1888 IV.–1889 XI.,

  p. 44 "Der Nöther'sche Satz (nach einer Mitteilung von Hilbert)"

  p. 45 "Hilberts Fundamentalsatz"

  p. 93 "Hilbert beweist die obigen Sätze so" (study on convergent series)

- No. 7: 1890 IV.9.–1891 XI.,

  p. 94 "[...] die Hilbert'schen Figuren" ("Lines on square"-figures)

- No. 8: 1891 XI.3.–1894 III.,

  p. 207 "Zweiter Hilbert'scher Formensatz" (Hilbert's basis theorem)

- No. 9: 1894 IV.4.–1895 I.6.,

  loose sheet concerning "[...] von Hilbert, betreffend die Anzahl von Covarianten"

- No. 13: 1895 VI.19.–XII.31.,

  p. 19 letter to Hilbert in stenography

- No. 14: 1896 I.1.–1897 II.1.,

  p. 204 "Hilberts 2tes Theorem" (related to [10, p. 485])

- No. 15: 1897 II.1.–1898 III.19.,

  p. 175 "Zu Hilberts "Körperbericht"" (related to Hilbert's *Zahlbericht*)

- No. 16: 1898 III.20.–1899 II.23.,

  p. 129 "Zum Hilbertschen Bericht pag. 287" (related to Hilbert's *Zahlbericht*)

- No. 19: 1901 XI.1.–1904 III.16.,

  p. 29 "Hilberts axiomatische Größenlehre" (related to [12])

  p. 114 "Abbildung einer Strecke auf ein Quadrat","[..] die Hilbert'schen geometrisch erklärten Funktionen sollen arithmetisch charakterisiert werden." ("Lines on square"-figures)

- No. 20: 1904 III.16.–1906 II.1.,

  p. 163 "Hilberts Beweis von Hadamards Determinantensatz"

- No. 21: 1906 II.1.–1906 XII.8.,

  p. 166 "Hilberts Vte Mitteilung über Integralgleichungen" (related to [14])

- No. 22: 1906 XII.18.–1908 I.22.,

  p. 36 "Convergenzsätze von Landau + Hilbert"

- No. 25: 1911 X.27.–1912 XII.27.,

  p. 77 "Zu Hilberts Formenarbeit (Charakt. Funktion eines Moduls)"

- In the directory HS 582: 32 due to George Pòlya on page 4 and 6 there are remarks "die ersten 9 Bände und Inhaltsverzeichnis sind zwecks Bearbeitung verderhand

bei Prof. Hilbert in Göttingen" and "22. zwecks Bearbeitung vorderhand bei Prof. Hilbert in Göttingen", crossed out with pencil
- HS 582: 28, Letter of condolences from David Hilbert to Ida Samuel-Hurwitz (December 15, 1919)
- HS 583: 52, Greeting cards from conferences with Hilbert's handwritting: "Lutetia Parisiorum, le 12 aout 1900" and the "Landau-Kommers 18. Jan. 1913"
- Remarks in the biographical dossier written by Ida Samuel-Hurwitz (HS 583a: 2).

Appendix II: Table of Figures

- Figure 1: Portraits of Adolf and Julius Hurwitz, taken from Riesz's register in *Acta Mathematica* from 1913 [42].
- Figure 2: Excerpt of [21, No. 9, p. 100], ETH Zurich library, Hs 582:9, DOI:10. 7891/e-manuscripta-12816.
- Figure 3: Excerpt of [30, p. 12].
- Figure 4: Illustration of Julius Hurwitz's types of partial quotients, made by the author.
- Figures 5, 6 and 7: Excerpts from the workshop notes, made by the author.
- Figure 8: Excerpt of [21, No. 6, p. 45], ETH Zurich library, Hs 582:6, DOI:10. 7891/e-manuscripta-12821.
- Figures 9 and 10: Excerpts of [21, No. 14] on pages 204 and 205, ETH Zurich library, Hs 582:14, DOI:10.7891/e-manuscripta-12840.
- Figure 11: Excerpt of [21, No. 15, p. 175], ETH Zurich library, Hs 582:15, DOI:10. 7891/e-manuscripta-12831.
- Figures 12 and 13: Excerpts of [21, No. 15, p. 177], ETH Zurich library, Hs 582:15, DOI:10.7891/e-manuscripta-12831.
- Figure 14: Excerpt of [21, No. 15, p. 178], ETH Zurich library, Hs 582:15, DOI:10. 7891/e-manuscripta-12831.
- Figures 15 and 17: Excerpts of [21, No. 16] on pages 129 and 130, ETH Zurich library, Hs 582:16, DOI:10.7891/e-manuscripta-12830.
- Figure 16: Excerpt of Hilbert's *'Zahlbericht'*, page 289 respectively [20, vl. I, p. 164].
- Figure 18: Excerpt of [21, No. 19, p. 29], ETH Zurich library, Hs 582:19, DOI:10. 7891/e-manuscripta-12819.
- Figure 19: Excerpt of [12, p. 182] and [21, No. 19, p. 29], ETH Zurich library, Hs 582:19, DOI:10.7891/e-manuscripta-12819.
- Figures 20 and 21: Excerpts of [21, No. 19, p. 30], ETH Zurich library, Hs 582:19, DOI:10.7891/e-manuscripta-12819.
- Figures 22, 23, 24, 25 and 26: Excerpts from [21, No. 21] on pages 166, 167, 168, 169 and 172, ETH Zurich library, Hs 582:21, DOI:10.7891/e-manuscripta-12836.
- Figures 27 and 28: Excerpts of [21, No. 18, p. 75, p. 81], ETH Zurich library, Hs 582:18, DOI:10.7891/e-manuscripta-12810.

- Figure 29: Greeting cards from "Lutetia Parisiorum, le 12 aout 1900" and the "Landau-Kommers 18. Jan. 1913", in Hs 583:53 and 57, ETH Zurich library.
- Figure 30: Excerpt of Georg Pòlya's list [21, No. 32, p. 6], ETH Zurich library, Hs 582:32, DOI:10.7891/e-manuscripta-16074.
- Figure 31: Illustration of Tanaka's change of coordinates $\{1, i\} \rightarrow \{\alpha, \overline{\alpha}\}$, made by the author.
- Figure 32: Illustration of the set of reciprocals X^{-1}, made by the author.
- Figure 33: Illustration of the numbers $a \in I$, made by the author.
- Figure 34: Illustration of the tiling of the complex plane in respect to the dual transformation, made by the author.

References

1. R. Bindel, in *Geschichte der höheren Lehranstalt in Quakenbrück.* Buchdruckerei von Heinrich Buddenberg (1904), https://archive.org/stream/geschichtederhh00bindgoogpage/n7/mode/2up
2. G.D. Birkhoff, Proof of the ergodic theorem. Proc. Nat. Acad. Sci USA **17**, 656–660 (1931)
3. O. Blumenthal, Lebensgeschichte. Gesammelte Abhandlungen von David Hilbert [20], Bd. **3**, 388–429 (1932)
4. W. Bosma, H. Jager, F. Wiedijk, Some metrical observations on the approximation by continued fractions. Indag. Math. **45**, 281–299 (1983)
5. W. Burau, Der Hamburger Mathematiker Hermann Schubert. Mitt. Math. Ges. Hamb. **9**, 10–19 (1966)
6. W. Burau, B. Renschuch, Ergänzungen zur Biographie von Hermann Schubert. Mitt. Math. Ges. Hamb. **13**, 63–65 (1993)
7. K. Dajani, C. Kraaikamp, *Ergodic Theory of Numbers* (Mathematical Association of America, Washington, 2002)
8. W. Doeblin, Remarques sur la théorie métriques des fractions continues. Compositio Math. **7**, 353–371 (1940)
9. J. Felker, L. Russell, High-precision entropy values for spanning trees in lattices. J. Phys. A **36**, 8361–8365 (2003)
10. D. Hilbert, Ueber die Theorie der algebraischen Formen. Math. Ann. **36**, 473–534 (1890)
11. D. Hilbert, Mathematische Probleme. Vortrag, gehalten auf dem internationalen Mathematiker-Kongreß zu Paris 1900. Gött. Nachr. **1900**, 253–297 (1900)
12. D. Hilbert, Über den Zahlbegriff. Jahresbericht der Deutschen Mathematiker-Vereinigung **8**, 180–183 (1900)
13. D. Hilbert, Grundzüge einer allgemeinen Theorie der linearen Integralgleichungen. Vierte Mitteilung. Gött. Nachr. **1904**, 49–91 (1904)
14. D. Hilbert, Grundzüge einer allgemeinen Theorie der linearen Integralgleichungen. Fünfte Mitteilung. Gött. Nachr. **1906**, 439–480 (1906)
15. D. Hilbert, Beweis für die Darstellbarkeit der ganzen Zahlen durch eine feste Anzahl n-ter Potenzen (Waringsches Problem). Math. Ann. **67**, 281–300 (1909)
16. D. Hilbert, Grundzüge einer allgemeinen Theorie der linearen Integralgleichungen. Sechste Mitteilung. Gött. Nachr. **1910**, 355–417 (1910)
17. D. Hilbert, Adolf Hurwitz. Math. Ann. **83**, 161–172 (1921)
18. D. Hilbert, *Grundlagen der Geometrie* B. G Teubner (1930)
19. D. Hilbert, Die Grundlegung der elementaren Zahlenlehre. Math. Ann. **104**, 485–494 (1931)
20. D. Hilbert, *Gesammelte Abhandlungen, 3 Bände.* Springer (1935)

21. A. Hurwitz, *Mathematische Tagebücher, in: Hurwitz, A., wissenschaftlicher Teilnachlass Hs 582, Hs 583, Hs 583a, ETH Zurich University Archives* (1882–1919), http://dx.doi.org/10.7891/e-manuscripta-12810 bis -12840

22. A. Hurwitz, Ueber die Entwicklung complexer Grössen in Kettenbrüche. Acta Math. **XI**, 187–200 (1888)

23. A. Hurwitz, Ueber eine besondere Art der Kettenbruchent-wickelung reeller Grössen. Acta Math. **XII**, 367–405 (1889)

24. A. Hurwitz, Review JFM 26.0235.01 of [30] in Jahrbuch über die Fortschritte der Mathematik, available via Zentralblatt (1894)

25. A. Hurwitz, Sur quelques applications geometriques des séries de Fourier. Ann. Ec. norm. Sup. **19**(3), 357 (1902)

26. A. Hurwitz, Über die Darstellung der ganzen Zahlen als Summe von n-ten Potenzen ganzer Zahlen. Math. Ann. **65**, 424–427 (1908)

27. A. Hurwitz, *Mathematische Werke. Bd. 1. Funktionentheorie; Bd. 2. Zahlentheorie, Algebra und Geometrie.* Birkhäuser, Basel (1932)

28. J. Hurwitz, Eberhard, Victor (1862–1927). Determinanten-Theorie nach Vorlesungen in Königsberg 1890/91 WS ausgearbeitet von Julius Hurwitz, in *Hurwitz, A., wissenschaftlicher Teilnachlass Hs 582 : 157, ETH Zurich University Archives* (1890)

29. J. Hurwitz, Hilbert, David (1862–1943). die eindeutigen Funktionen mit linearen Transformationen in sich, nach Vorlesungen in Königsberg 1892 SS ausgearbeitet von Julius Hurwitz. in *Hurwitz, A., wissenschaftlicher Teilnachlass Hs 582 : 154, ETH Zurich University Archives* (1892)

30. J. Hurwitz, Ueber eine besondere Art der Kettenbruchent-wicklung complexer Grössen, Dissertation, University of Halle. printed by Ehrhardt Karras in Halle (1895)

31. J. Hurwitz, Über die Reduktion der binären quadratischen Formen mit komplexen Koeffizienten und Variabeln. Acta Math. **25**, 231–290 (1902)

32. P. Kasteleyn, The statistics of dimmer on a lattice. Physica **27**, 1209–1225 (1961)

33. A. Khintchine, Zu Birkhoffs Lösung des Ergodenproblems. Math. Ann. **107**, 485–488 (1933)

34. H. Nakada, The metrical theory of complex continued fractions. Acta Arith. **LVI**, 279–289 (1990)

35. N. Oswald, David Hilbert, ein Schüler von Adolf Hurwitz? Siegener Beiträge zur Geschichte und Philosophie der Mathematik **4**, 11–30 (2014)

36. N. Oswald, Hurwitz's Complex Continued Fractions. A Historical Approach and Modern Perspectives. (Dissertation) OPUS, Online-Publikationsservice der Universität Würzburg (2014)

37. N. Oswald, The Doeblin-Lenstra conjecture for a complex continued fraction algorithm. Tokyo J. Math. **39**, (in print) (2016)

38. N. Oswald, J. Steuding, Complex Continued Fractions - Early Work of the Brothers Adolf and Julius Hurwitz. Arch. Hist. Exact Sci. **68**, 499–528 (2014)

39. G. Pólya, *The Polya Picture Album: Encounters of a Mathematician editen by G.L. Alexanderson* (Birkhäuser, 1987)

40. C. Reid, *Hilbert* (Springer, Berlin, 1970)

41. B. Renschuch, Zur Definition der Grundideale. Mathematische Nachrichten **55**, 63–71 (1973)

42. M. Riesz, *Acta mathematica, 1882-1912. Table générale des tomes 1–35* (Almqvist and Wiksells, Uppsala, 1913)

43. D. Rowe, "Jewish Mathematics" at Göttingen in the Era of Felix Klein. Chic. J. ISIS **77**, 422–449 (1986)

44. D. Rowe, Felix Klein, Adolf Hurwitz, and the "Jewish question" in German academia. Math. Intell. **29**, 18–30 (2007)

45. I. Samuel-Hurwitz, *Erinnerungen an die Familie Hurwitz, mit Biographie ihres Gatten Adolph Hurwitz*, Prof. f. höhere Mathematik an der ETH, Zürich, ETH library (1984)

46. S. Tanaka, A complex continued fraction transformation and its ergodic properties. Tokyo J. Math. **8**, 191–214 (1985)

47. C. Tapp, *An den Grenzen des Endlichen - Das Hilbertprogramm im Kontext von Formalismus und Finitismus* (Springer - Spektrum, 2013)

48. H. von Mangoldt, Briefe von Hans von Mangoldt an Adolf Hurwitz. *Correspondence of Adolf Hurwitz, Cod Ms Math Arch 78, letter 124, Niedersächsische tSaats- und Universitätsbibliothek Göttingen* (1884)

49. A. Wangerin, Gutachten Julius Hurwitz. *Universitätsarchiv Halle, Rep. 21 Nr. 162* (1895)

Index

© Springer International Publishing AG 2016
J. Steuding (ed.), *Diophantine Analysis*, Trends in Mathematics,
DOI 10.1007/978-3-319-48817-2